A PRACTICAL FIELD GUIDE FOR AS9100C

Also available from ASQ Quality Press:

A Practical Field Guide for ISO 9001:2008
Erik Valdemar Myhrberg

ISO 9001:2008 Explained, Third Edition
Charles A. Cianfrani, John E. "Jack" West, and Joseph J. Tsiakals

ISO Lesson Guide 2008: Pocket Guide to ISO 9001-2008, Third Edition
J.P. Russell and Dennis R. Arter

ISO 9001:2008 Internal Audits Made Easy: Tools, Techniques, and Step-By-Step Guidelines for Successful Internal Audits, Second Edition
Ann W. Phillips

Process Driven Comprehensive Auditing: A New Way to Conduct ISO 9001:2008 Internal Audits, Second Edition
Paul C. Palmes

ISO 9001:2008 Interpretive Guide for the Design and Construction Project Team (e-Book)
Prepared by members of the ASQ Design and Construction Division and edited by John R. Broomfield

How to Audit the Process-Based QMS
Dennis R. Arter, John E. (Jack) West, and Charles A. Cianfrani

The ASQ Auditing Handbook, Third Edition
J.P. Russell, editing director

Quality Audits for Improved Performance, Third Edition
Dennis R. Arter

The Quality Toolbox, Second Edition
Nancy R. Tague

Mapping Work Processes, Second Edition
Bjørn Andersen, Tom Fagerhaug, Bjørnar Henriksen, and Lars E. Onsøyen

Lean Kaizen: A Simplified Approach to Process Improvements
George Alukal and Anthony Manos

Root Cause Analysis: Simplified Tools and Techniques, Second Edition
Bjørn Andersen and Tom Fagerhaug

The Certified Manager of Quality/Organizational Excellence Handbook, Third Edition
Russell T. Westcott, editor

A PRACTICAL FIELD GUIDE FOR AS9100C

Erik Valdemar Myhrberg
with
Dawn Holly Crabtree and Rudolph "RE" Hacker

ASQ Quality Press
Milwaukee, Wisconsin

American Society for Quality, Quality Press, Milwaukee 53203

Printed in the United States of America
18 17 16 15 14 5 4 3

Library of Congress Cataloging-in-Publication Data

Myhrberg, Erik V.
A practical field guide for AS9100C / Erik Valdemar Myhrberg ; with Dawn Holly Crabtree and Rudolph "RE" Hacker.
p. cm.
Includes bibliographical references.
ISBN 978-0-87389-793-8 (alk. paper)
1. Aerospace engineering—Quality control—Standards. 2. Aerospace industries—Management. 3. ISO 9000 Series Standards. I. Crabtree, Dawn Holly. II. Hacker, Rudolph. III. Title.

TL671.28.M94 2010
629.1068'4—dc22 2010021511

ISBN: 978-0-87389-793-8

Publisher: William A. Tony
Acquisitions Editor: Matt T. Meinholz
Project Editor: Paul O'Mara
Production Administrator: Randall Benson

ASQ Mission: The American Society for Quality advances individual, organizational, and community excellence worldwide through learning, quality improvement, and knowledge exchange.

Attention Bookstores, Wholesalers, Schools, and Corporations: ASQ Quality Press books, video, audio, and software are available at quantity discounts with bulk purchases for business, educational, or instructional use. For information, please contact ASQ Quality Press at 800-248-1946, or write to ASQ Quality Press, P.O. Box 3005, Milwaukee, WI 53201-3005.

To place orders or to request ASQ membership information, call 800-248-1946. Visit our Web site at http://www.asq.org/quality-press.

 Printed on acid-free paper

Quality Press
600 N. Plankinton Avenue
Milwaukee, Wisconsin 53203
Call toll free 800-248-1946
Fax 414-272-1734
www.asq.org
http://www.asq.org/quality-press
http://standardsgroup.asq.org
E-mail: authors@asq.org

We dedicate this combined work
to our loving families.

CONTENTS

PREFACE

Now solidly in the twenty-first century, companies are being pushed to the extremes of their limited resources. At one time it was sufficient to meet most of our customer's requirements, but not now. On a global scale, we are all being asked to be better, faster, and cheaper than our competitors or our customers will look elsewhere. At some point soon the current internal processes will not be able to hold back the deluge, and companies will be faced with a stark decision—consistently improve or perish.

Companies, teams, and individuals are still trying to meet customer expectations in an effective and efficient manner, worldwide competition still drives the need for innovation, and internal process pressures still demand continual improvement in order to remain functional. One of the best and most widely accepted ways in which companies can face these challenges is to implement an effective *and* efficient quality management system, which not only adds value to the organization but also satisfies the customer.

It is extremely challenging to stay competitive in the aerospace industry; the need for assurance of good material is paramount and has driven cost, sometimes unnecessarily due to the continued existence of outdated manufacturing and quality system thinking and practices. Cost is primarily due to small order quantities, product complexity, and special processes and materials. Yet, traditional production, over-inspection, and processing techniques are also extremely costly, as typically many redundant tests of the same feature occur to assure conformance.

When MIL-Q-9858 was retired, the aerospace industry started to feel the ill effects of the absence of aerospace-specific mandated criteria. Though many companies registered to ISO 9001, which provided a fantastic yet generic business quality management system, it did not address the needs of the aerospace industry. To those of us who could see the wastes in the traditional "product realization" system, the creation of AS9000, and subsequently AS9100 based on ISO 9001, was by divine providence. Here was a QMS that focused on process control that is efficient and effective, not redundant inspection, and communicated the industry-specific needs of aerospace—such as first inspection and product traceability. When ISO 9001 became a process-oriented standard in the year 2000 it better supported the process control structure of AS9100; hence, AS9100:2001 is the best QMS standard for aerospace—bar none!

From working with and within aerospace primes, medium-size suppliers, job shops, and software design, it has become apparent how little the concept of process control has caught on. Combining the experience of over 150 QMS implementations with responsibilities covering internal and supplier quality engineering, and training and implementing of lean manufacturing and Six Sigma quality, we readily see how businesses could avoid long-term costs of redundant non-value-adding tasks like inspections, scrap, rework, and customer returns by following the requirements of AS9100 using proven quality tools. AS9100 is not prescriptive, but gives hints as to the use of these tools.

The purpose of this field guide is to assist the reader, step by step, in implementing a QMS in conformance with AS9100. This field guide has been created in order to facilitate an inner reliance between senior management, middle management, functional teams, and the individual. Users of the field guide will find within it practical tools, tips, and techniques useful for not only implementing a quality management system but also for maintaining one.

In keeping with ISO 9000's definition of "quality" as the "degree to which a set of inherent characteristics fulfils requirements," we have identified the requirements and inherent characteristics (distinguishing features) of AS9100. Within this field guide, each subclause containing requirements is the focus of a two-page spread that consistently presents the features listed below.

Provisions (or What the Field Guide Will Do)

- Provide a user-friendly guide to AS9100's requirements for implementation purposes
- Identify the documents/documentation required, along with recommendations on what to consider retaining/adding to a QMS during AS9100 implementation
- Guide internal auditor(s) regarding what to ask to verify that a conforming and effective QMS exists
- Direct management on what it must do and should consider to satisfy AS9100's enhanced requirements and responsibilities for top management
- Depict step by step what must occur to create an effective, conforming QMS

Direct Characteristics (or What the Field Guide Provides)

- *The standard*—A paraphrase of what a subclause of AS9100 requires in easy-to-understand language, with references to information in ISO 9000 and guidance in ISO 9004 to enhance the user's understanding of what AS9100 requires and what possible added steps the user can take to improve performance (per the ISO 9004 guidelines)
- *Document requirements*—A list of the documentation/documents required by the subclause, with some ideas to consider in satisfying those requirements that will take the system beyond the requirements toward continual improvement
- *Internal audit questions*—What every internal audit team needs to ask at a minimum when assessing the QMS for conformance with the subclause

- *Management summary*—A concise description of what management must do and/or is responsible for in order to achieve conformance to the subclause, along with some guidance on additional steps management can take to enhance the system
- *Process control tip*—Recommendations of method or tools to best infuse process control into the QMS, as AS9100 was intended to be used
- *Subclause flowchart*—A depiction of the steps that need to be undertaken during an implementation/transition effort to effectively and efficiently satisfy the requirements of the subclause of AS9100, along with a box providing guidance on use of the flowchart

This field guide is designed to provide you with a consistent approach to implementing an AS9100-conforming QMS, which is appropriate since AS9100 continues to view quality as the ability of an organization to consistently deliver product that meets customer specifications. The field guide examines each subclause of Sections 4–8 of AS9100, which contain the requirements.

Subclause 1.2, Scope—Application, does not contain requirements but is critical to properly excluding any requirements of AS9100 that do not apply to your organization's QMS and should therefore be treated as an important part of the field guide due to the importance of establishing the QMS's scope.

Paragraphs and items in *bold italics* delineate AS9100 requirements above and beyond those of ISO 9001. This includes Clauses, Internal Audit Questions, Management Summary, Process Control Tips, and Flowcharts.

What separates this field guide from most other books on AS9100 and its implementation are the flowcharts showing the steps to be taken in implementing a QMS to meet a subclause's requirements. But the flowcharts themselves can be overwhelming when you first look at them. For this reason, a box appears with each flowchart that explains pertinent facts and/or what the flowchart represents and how it is to be used.

Remember, the QMS your organization implements must meet the needs of its users—you and the rest of your organization's employees, from senior management to the most junior employee. So the QMS you create using this field guide will not only satisfy AS9100's requirements, but provide a set of processes that suits your organization and fosters improved use of the system and improvement in the processes of the organization over time. You may also use the ISO 9004 guidelines for further explanation. Please note that compliance to AS9100 may not ensure that your organization meets all contractual requirements.

A goal throughout this field guide is to provide clear concepts of how to ingrain process control into your business utilizing the AS9100 system. Your goal: for your processes to turn out aerospace conforming parts on time, every time, at a competitive cost. The companies that can compete now and in the future will need to do just that. Our hope is that our AS9100 field guide will help to get you there.

To your continued success,
Erik Myhrberg with Dawn Crabtree and Rudy Hacker

SECTION 4: QUALITY MANAGEMENT SYSTEM

4.0 Quality Management System

4.1 General Requirements

AS9100 The organization must develop, document, set up, and maintain a QMS and continually improve its effectiveness through six activities, including (1) identification of needed processes, (2) making sure adequate resources and (3) information are available to support both the processes and their (4) monitoring, and (5) achieving expected results and (6) continual improvement of the processes though implemented actions. All related activities must conform to ISO 9001's requirements. When outsourcing related to product occurs, the QMS must contain provisions to ensure the organization controls the processes involved. A Note indicates processes that should be included in the QMS.

ISO 9004 Offers guidance on activities top management should pursue to create a customer-oriented organization and move it and its QMS toward continual improvement and improved performance. Annexes A and B provide guidance on self-assessment and process improvement.

Document Requirements:

Required:

No specific documents, although this subclause relates to and affects other requirements in ISO 9001:2008 that concern specific QMS documents/documentation.

Remember:

- Processes need to be clearly defined, with the sequence and interactions between stages and with other processes defined. This will require documenting the organization's operations.
- The defined processes must include all administration and management activities.

Internal Audit Questions:	*Management Summary:*
• Has the quality management system been established, documented, implemented, and maintained? • Have processes that are necessary for maintaining the quality management system been determined? • Has the organization demonstrated continual improvement? • Does the organization have control over processes that are outsourced?	Top management must: • Define systems and processes • Ensure that they are trained and managed • Ensure effective control of processes • Utilize data to evaluate performance Top management must follow these quality management system principles: • Have a true customer focus • Demonstrate leadership • Involve people • Use a process-oriented approach • Use a systems approach to management • Strive for continual improvement • Base decisions on facts and data • Cultivate mutually beneficial supplier relationships ***Process Control Tip:*** ***Process mapping the current state of your business will help to identify inconsistencies in operations and improvement opportunities. Involve employees that supply, perform, and are customers to the process to participate in creation of process maps. Since they define your business methods, these maps should be the main content in your procedures.*** ***Process maps provide a visual yet comprehensive view of the process and ease in changing procedures and in training new employees.***

4.1 General Requirements

To leverage your QMS for greater improvement you need to:

- Establish
- Document
- Implement
- Maintain
- Continually improve

your quality management system.

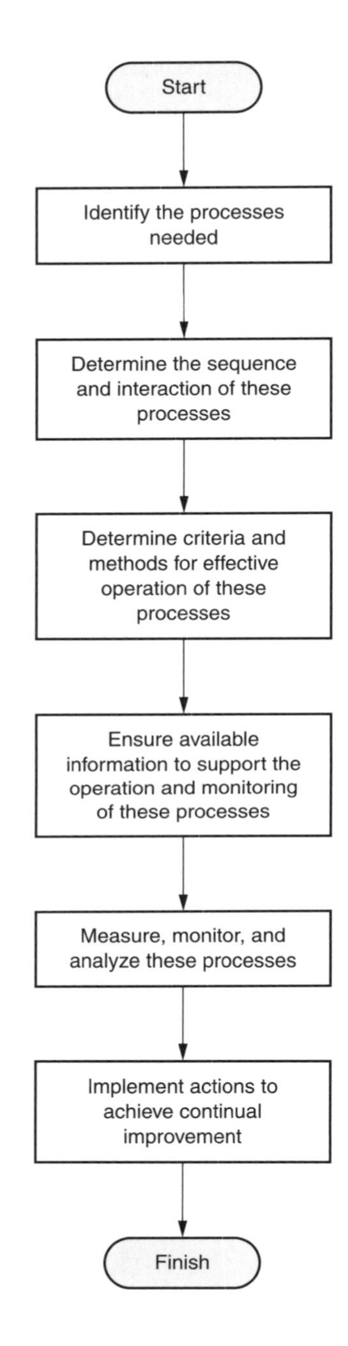

4.0 Quality Management System

4.2 Documentation Requirements

4.2.1 General

AS9100 Specifies what is to be included in the QMS documentation: quality policy, objectives, and manual; procedures that must be documented to conform to ISO 9001; documents required for "effective planning, operation, and control" of the organization's processes; quality records; ***all customer and regulatory requirements.*** Three Notes define what is meant by "documented procedure," clarify that the amount and level of QMS documentation will vary among organizations for three reasons, and explain that documentation can be in any form or medium (for example, flowchart, electronic file).

ISO 9004 Suggests management should define the required QMS documentation for the organization, explains what the documentation should do, and provides factors management should consider in defining the QMS documentation.

ISO 9000 Defines a "document" in 3.7.2 as "information and its supporting medium" and provides examples of types of documents. Explores in 2.7, Documentation, the value of documentation and the types of documents used in a QMS.

4.2.2 Quality Manual

AS9100 A quality manual is to be developed and maintained that includes the QMS's scope—with full explanation for why any requirements are excluded—its documented procedures (or indications of where to find them), and a description of the process approach as it works in the QMS.

ISO 9000 Defines a "quality manual" in 3.7.4 as a "document specifying the [QMS] of an organization."

Document Requirements:

Remember:

- A quality policy (statement *and* objectives), quality manual, the procedures specified, documents your organization needs to function and meet customer/regulatory requirements and records conformance.
- The notes in this subclause section can have a significant bearing on the documentation when establishing the QMS.

Internal Audit Questions:

- Does the quality management system include documented statements of the quality policy and quality objectives?
- Is there a quality manual?
- Are the required documented procedures in place?
- Are there additional documented procedures where determined as needed by the organization?
- Is the quality management system documentation based on the size, type, complexity, and interaction of processes in the organization?
- Are the required records available?
- *Can you readily identify which requirement(s) are addressed in each procedure?*

Management Summary:

Management must define the documentation, including records, that:

- Are required to establish, implement, and maintain a QMS
- Support an effective and efficient operation

Process Control Tip:

Matrices of customer and regulatory requirements referenced to your QMS will assist in assuring compliance, during creation and change of processes, and greatly assists auditors.

4.2 Documentation Requirements of a Quality Management System

4.2.1 General

4.2.2 The Quality Manual

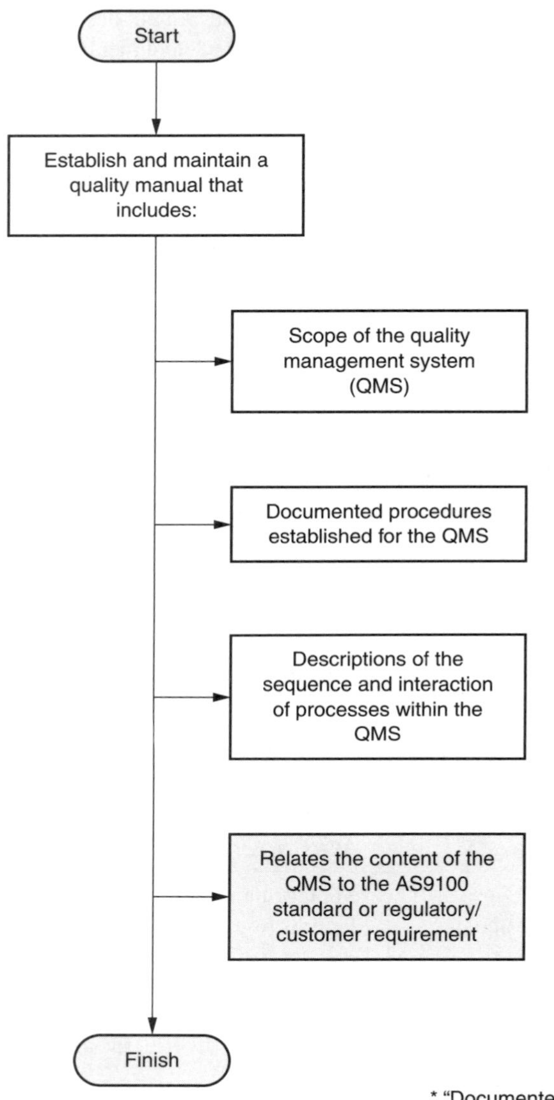

* "Documented Procedures" must be:

- Established
- Documented
- Implemented
- Maintained

4.0 Quality Management System

4.2 Documentation Requirements

4.2.3 Control of Documents

AS9100 The organization must develop and document a procedure that defines the necessary controls to ensure that eight document-related processes are established and provide the level of document control required by ISO 9001. These processes include document approval, review, updating and reapproval, revision status identification, availability of documents where they are needed, legibility, identification and control of external documents, prevention of unintended use of obsolete documents, and coordination of document changes with customer and/or regulatory authorities in accordance with contract or regulatory requirements.

ISO 9004 Advises that control of documentation should be evaluated to ensure that the procedures support the organization's effectiveness and efficiency when seven criteria are considered.

Document Requirements:

Required:

Document control procedure

Remember:

The review, updating, and reapproval process has always been a requirement; however, you must now add the concept that the documents themselves must be reviewed.

Internal Audit Questions:

- Is document control established in a documented procedure?
- Are documents reviewed, updated, and approved for adequacy prior to use? Are document changes reapproved to ensure adequacy prior to use? Is current document-revision status maintained?
- Are obsolete documents that are retained for any purpose suitably identified to prevent unintended use?
- Is there a process to ensure that documents remain legible, readily identifiable, and retrievable?
- Are latest versions of applicable documents available at points of use? Are relevant versions of external documents identified and controlled?

Management Summary:

Management should:

- Define the documentation required
- Give consideration to contractual requirements, acceptance of standards, regulatory requirements, organizational decisions, needs and expectations of interested parties
- Evaluate the generation, use, and control of documentation against criteria such as functionality, user-friendliness, resources, policies and objectives, managing knowledge, benchmarking documentation, and interfaces used by outside sources
- Feed back proposed document changes to contract review to ensure compliance with contracts and notification to customer and/or regulatory authorities where required

Process Control Tip:

A control plan for any process addresses change management and, where required, should list applicable contractual requirements. This ensures that during process changes customer and regulatory requirements are always adhered to.

4.2 Documentation Requirements

4.2.3 Control of Documents

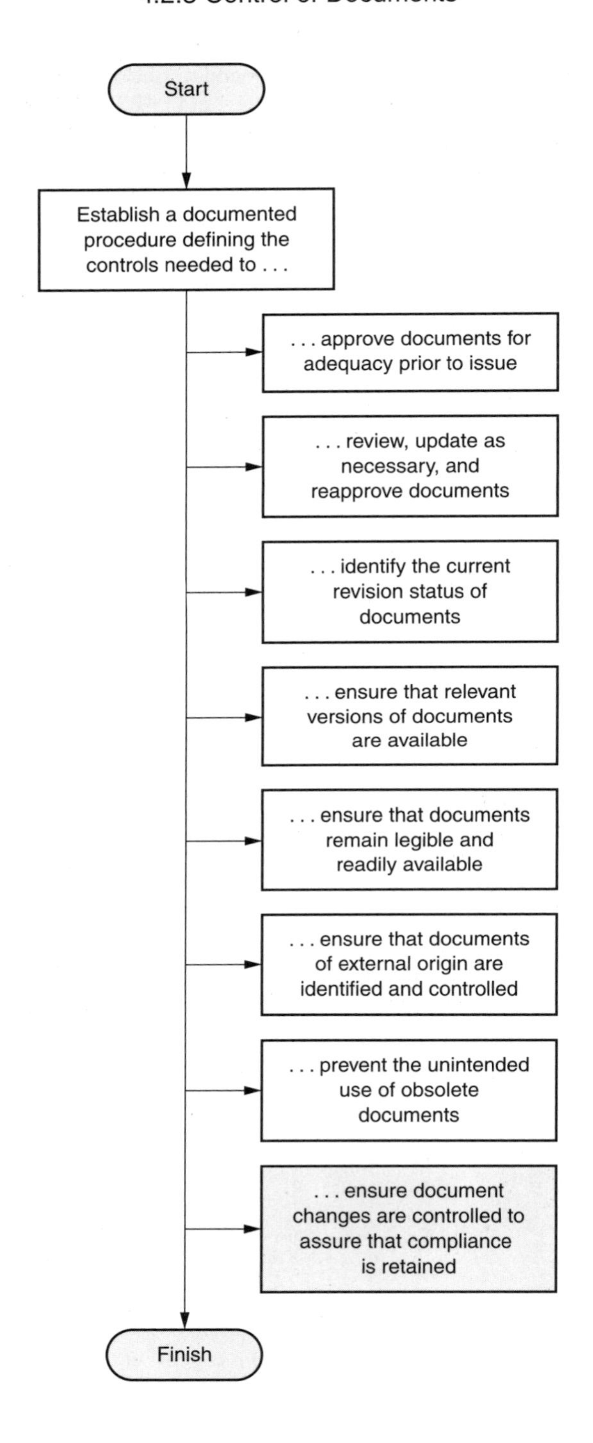

4.0 Quality Management System

4.2 Documentation Requirements

4.2.4 Control of Records

AS9100 Quality records are to be "established and maintained" to prove that the QMS is operating as intended and is effective and satisfies the requirements of ISO 9001. The organization must establish a specific procedure in addition to those for 4.2.3 to define the controls that will apply to quality records and specify, among other things, their storage, retrieval, retention time, and disposition; ***this includes controlling quality records retained by your supplier.***

ISO 9000 Defines a "record" in 3.7.6 as a "document stating results achieved or providing evidence of activities performed."

Document Requirements:

Required:

Documented procedure

Remember:

The procedure must not only control records, but do so in a way that serves the organization's needs without creating valueless paperwork and bureaucracy.

Internal Audit Questions:

- Is there a documented procedure for control of records?
- Have the organization's records been catalogued?
- Are protection requirements in place for all records?
- Have retention times and disposition requirements been established for all records?
- Are records disposed of as required by the established procedures?
- Have storage and retrieval requirements been determined and implemented for all records?
- Is there a documented procedure addressing supplier control?
- Are you flowing down to suppliers the requirement to control documents where the organization has chosen for the supplier to control them?
- Have you verified that your suppliers are controlling applicable documentation?

Management Summary:

Management must establish and maintain records in order to provide objective evidence of conformance to the organization's QMS.

Records retained by suppliers shall be controlled and made available on request; therefore, supplier contracts (purchase orders) must flow down this requirement (refer to Section 7.4).

All quality records shall be made available to customer or regulatory authorities per contractual requirements.

Process Control Tip:

1. ***Implement a visual document system for simplification and easy retrieval.***
2. ***Create a document control plan that links to contact requirements; this ensures the proper availability of records that may be requested by customer/regulatory authorities (see Process Control Tip in element 4.2.1).***

4.2 Documentation Requirements

4.2.4 Control of Records

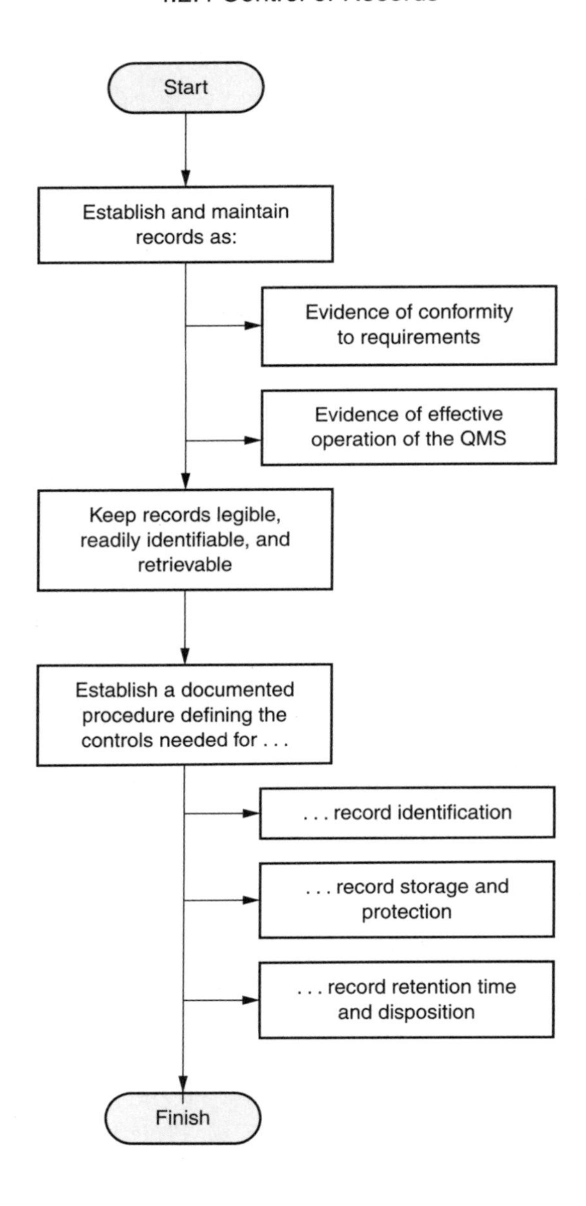

4.0 Quality Management System

4.3 Configuration Management

AS9100 ***A controlled process of configuration management for all product is required to assure best quality in design and control of the design.***

ISO10007 ***Guide to Configuration Management***

This document will simplify the design and implementation of a robust configuration management system.

Document Requirements:

Required:

Documented procedure

Remember:

- The procedure should simplify the configuration process.
- All design and contract reviews should feed into this process.

Internal Audit Questions:

- Is there a documented procedure for configuration management?
- Are configurations current?
- Is the configuration management process appropriate for the product?

Management Summary:

Process Control Tip:

Develop and map a robust process to ensure that all feedback from suppliers, customers, and internal processes that recommend design or modification to product are controlled and the best use of the change control process is utilized.

REVISIONS TO AS9100:2009 (REVISION C), CLAUSE 4

(Source: Larry Whittington)

4. Quality Management System

4.1 General Requirements

AS9100C inserts the following sentence to add customer and legal requirements to those of AS9100C. The rationale for this addition is to indicate that these requirements apply at the quality management system level, not just at the previously stated documentation level:

> *The organization's quality management system shall also address customer and applicable statutory and regulatory quality management system requirements.*

In 4.1, subclause (a), the word "Identify" has been replaced with "Determine":

> *a) ~~Identify~~ determine the processes needed for the quality management system and their application throughout the organization (see 1.2),*

Although similar, the words "Identify" and "Determine" have slightly different meanings. To *identify* is to recognize or establish something as being a particular thing. To *determine* is to apply reason and reach a decision. To determine the processes implies more analysis and judgment than merely identifying them.

> *e) monitor, measure where applicable, and analyze these processes, and . . .*

Processes are monitored, but may not need to be measured. Therefore, the requirement change above indicates that processes are only measured where applicable.

Later in clause 4.1, regarding outsourcing:

> *Where an organization chooses to outsource any process that affects product conformity ~~with~~ to requirements, the organization shall ensure control over such processes. ~~Control of such~~ The type and extent of control to be applied to these outsourced processes shall be ~~identified~~ defined within the quality management system.*

This addition clarifies that specific controls are to be defined and applied, not just identified. See the new Note 3 below for an explanation of the type and extent of controls for an outsourced process.

The current Note under clause 4.1 has been expanded, and two new Notes have been added:

> *NOTE 1: Processes needed for the quality management system referred to above ~~should~~ include processes for management activities, provision of resources, product realization, ~~and~~ measurement, analysis, and improvement.*

The text above expands from "measurement" to "measurement, analysis, and improvement" to match the title for clause 8. And, by deleting "should," it clearly states that these processes are included.

The new Note below provides an explanation of what is considered an outsourced process:

> *NOTE 1: Processes needed for the quality management system referred to above ~~should~~ include processes for management activities, provision of resources, product realization, ~~and~~ measurement, analysis, and improvement.*

> *NOTE 2: An outsourced process is a process that the organization needs for its quality management system and which the organization chooses to have performed by an external party.*

The new Note below identifies the factors influencing the control of an outsourced process:

> *NOTE 3: Ensuring control over outsourced processes does not absolve the organization of the responsibility of conformity to all customer, statutory, and regulatory requirements. The type and extent of control to be applied to the outsourced process can be influenced by factors such as*
>
> *a) the potential impact of the outsourced process on the organization's capability to provide product that conforms to requirements;*
>
> *b) the degree to which the control for the process is shared;*
>
> *c) the capability of achieving the necessary control through the application of clause 7.4.*

Outsourcing a process to another organization typically involves the purchase of those services. As a result, the requirements of clause 7.4, including the controls mentioned in 7.4.1, apply equally to the supplier selected to perform the outsourced process.

4.2 Documentation Requirements

4.2.1 General

The changes in this section are basically just a restructuring of the subclauses c), d), and e):

> *c) documented procedures and records required by this International Standard, and*
>
> *d) documents, including records, ~~needed~~ determined by the organization to be necessary to ensure the effective planning, operation, and control of its processes. ~~, and~~*
>
> *~~e) records required by this International Standard (see 4.2.4).~~*

You can see that adding "records" to subclause (c) allowed subclause (e) to be dropped. Subclause (d) has been expanded to include the necessary records.

AS9100C removed subclause (f) because its new general requirement in clause 4.1 more broadly states that the system must address the applicable statutory and regulatory quality management system requirements:

> *~~f) quality system requirements imposed by the applicable regulatory authorities.~~*

AS9100C reworded the first sentence of this paragraph and deleted the second sentence. It makes sense that personnel not just have access to, but be made aware of the documentation and changes:

> *The organization shall ensure that personnel have access to, and are aware of, relevant quality management system documentation and changes. ~~are aware of relevant procedures.~~ ~~Customer and/or regulatory authorities' representatives shall have access to quality management system documentation.~~*

The first Note for clause 4.2.1 has added two more sentences:

> *A single document may include the requirements for one or more procedures. A requirement for a documented procedure may be covered by more than one document.*

An example for the first sentence would be to satisfy the requirements for documented procedures in 8.5.2, Corrective Action, and 8.5.3, Preventive Action, through one combined corrective and preventive action procedure. An example for the second sentence would be to split the required procedure for the control of documents into two separate documented procedures.

4.2.2 Quality Manual

AS9100C deleted this entry under 4.2.2, subclause (b) regarding the relationship between requirements and documented procedures. It was too prescriptive and didn't really contribute to product quality:

> *~~when referencing the documented procedures, the relationship between the requirements of this International Standard and the documented procedures shall be clearly shown.~~*

4.2.3 Control of Documents

The opening sentence of this clause in ISO 9001:2008 still states that documents required by the quality management system are to be controlled. The only revision to clause 4.2.3 is shown below:

> *f) to ensure that documents of external origin determined by the organization to be necessary for the planning and operation of the quality management system are identified and their distribution controlled, and*

The change in subclause (f) clarifies that not all external documents have to be identified and controlled—only those needed for the planning and operation of the quality management system.

AS9100C deleted this sentence because the 4.1 addition in General Requirements addresses customer and legal requirements for the entire system:

> *~~The organization shall coordinate document changes with customers and/or regulatory authorities in accordance with contract or regulatory requirements.~~*

4.2.4 Control of Records

The opening sentence of clause 4.2.4 has expanded from records being "maintained" to having them "controlled." Maintaining records would simply keep them in good condition. Controlling the records means to regulate their use:

> *Records ~~shall be~~ established ~~and maintained~~ to provide evidence of conformity to requirements and of the effective operation of the quality management system shall be controlled.*
>
> *~~Records shall remain legible, readily identifiable, and retrievable.~~*
>
> *The organization shall establish a documented procedure ~~shall be established~~ to define the controls needed for the identification, storage, protection, retrieval, retention ~~time~~, and disposition of records.*
>
> *Records shall remain legible, readily identifiable, and retrievable.*

The requirement for a documented record control procedure has been rewritten, but the content is basically the same. It is now a separate paragraph for emphasis and moved up in the section.

Note that "retention time" has been reduced to "retention." And, you can see that records must still remain legible, readily identifiable, and retrievable. This requirement is now a separate paragraph and moved to the end of clause 4.2.4.

AS9100C deleted the following sentence from clause 4.2.4 because the 4.1 addition in General Requirements addresses customer and legal requirements for the entire system:

~~*Records shall be available for review by customers and regulatory authorities in accordance with contract or regulatory requirements.*~~

The Configuration Management clause 4.3 in AS9100B has been moved to clause 7.1.3 in AS9100C:

~~*4.3 Configuration Management*~~

SECTION 5: MANAGEMENT RESPONSIBILITY

5.1— Management Commitment
5.2—Customer Focus
5.3—Quality Policy
5.4—Planning
5.5—Responsibility, Authority, and Communication
5.6—Management Review

5.0 Management Responsibility

5.1 Management Commitment

AS9100	Top management must be committed to creation and implementation of a QMS and its continual improvement and will demonstrate its commitment through five types of activities. These activities include organizationwide communication of the criticality of meeting customer and other requirements, establishment of the quality policy and objectives, management reviews, and making sure QMS resources are available.
ISO 9004	Suggests actions top management should consider to maximize customer satisfaction and achieve benefits for all parties, recommends methods of performance measurement to help achieve objectives, explores activities relating to the quality management principles that top management should "demonstrate leadership in, and commitment to," and notes activities that should be considered to optimize processes.
ISO 9000	Discusses in 2.6 Role of Top Management within the Quality Management System.

Document Requirements:

Required:

No specific documents, although evidence of top management's commitment to the QMS must be present

Remember:

- This clause is a reminder to top management that there must be a process not only to *create* awareness of the organization's quality policy and quality objectives, but also to *maintain* this awareness.
- In addition, top management has enhanced responsibilities in its role with ISO 9001:2008 and the need to demonstrate its commitment to the QMS in ways that can be not just audited by auditors but used by the rest of the organization to maintain and improve processes.

Internal Audit Questions:

- Has management established quality objectives and developed a quality policy?
- Are management reviews performed?
- Does management provide and review the adequacy of resources?
- Is a process in place to ensure that all employees understand the importance of fulfilling customer, regulatory, and legal requirements?
- Is a process in place to develop specific programs for communicating customer and regulatory requirements?
- Is there objective evidence of conformance to these programs in management review records?

Management Summary:

Top management should strive to be a model of leadership, show commitment, and be actively involved in achieving customer satisfaction and regulatory and legal compliance.

Top management should establish the quality policy as a sign of its commitment, and the policy should drive continual improvement of the QMS's effectiveness.

Top management must define measurement methods to determine if planned objectives have been achieved.

When developing, implementing, and managing the QMS, top management must be involved in the development of quality objectives, either by participating in their setting or by providing the framework and environment for their creation.

Top management must evaluate the QMS regularly.

When developing, implementing, and managing the QMS, top management must consider the principles listed in the Management Summary for clause 4.1.

5.1 Management Commitment

These are the actions required of the organization's top management.

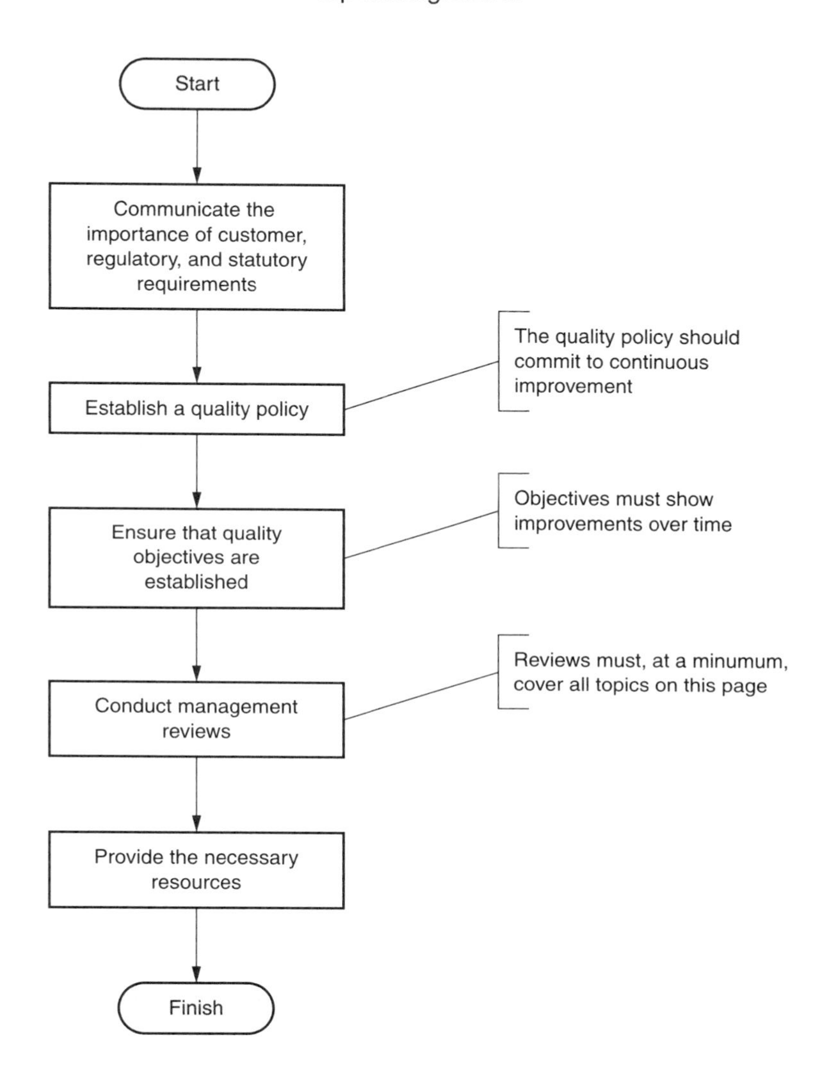

5.0 Management Responsibility

5.2 Customer Focus

AS9100 To increase customer satisfaction, top management must ensure that the organization understands and is capable of satisfying customer requirements (see 7.2.1 and 8.2.1).

ISO 9004 Provides examples of customer/end user needs and guidance on what an organization should do to understand and satisfy the needs of all interested parties. Explores the value of partnerships and what an organization should undertake in regard to its relationship with society.

Document Requirements:

Remember:

Top management must be able to demonstrate that they have put in place processes to make certain that these requirements are met.

Internal Audit Questions:

- Is there a process in place that precisely determines customer needs and expectations?
- Is there a process in place to convert customer needs and expectations into company-specific requirements?
- What objective evidence is available to confirm that these processes were used?
- How has top management led the organization in such a way that the organization's focus is on meeting the needs of the customers, not just making a product?

Management Summary:

The organization should identify personnel needs and expectations for recognition, work satisfaction, and personal development in order to ensure the involvement and motivation of its people with relation to customer satisfaction.

In order to understand and satisfy the current and future needs of customers, the organization should:

- Identify, assess, and respond to customer needs and expectations
- Translate customer needs and expectations into requirements
- Communicate the requirements throughout the organization
- Focus on process improvements to ensure satisfaction

5.2 Customer Focus

Top management must demonstrate a strong commitment to customer satisfaction

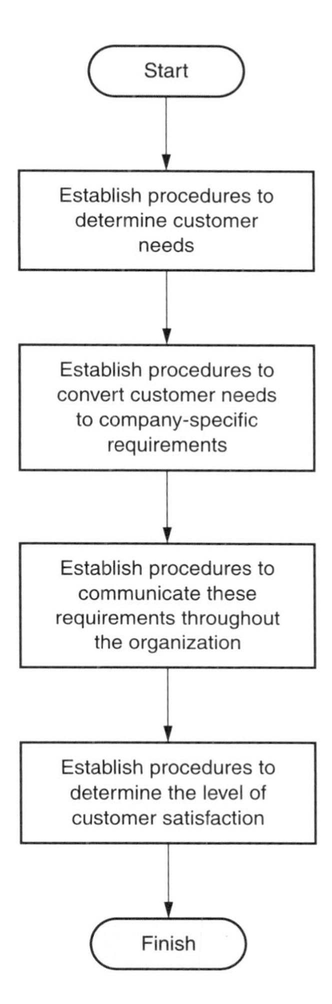

5.0 Management Responsibility

5.3 Quality Policy

AS9100 Top management is responsible for guaranteeing that the quality policy is appropriate, commits the organization to conformance with all requirements and to continual improvement of the QMS, is a suitable basis for quality objectives, is disseminated and understandable to the entire organization, and is reviewed for ongoing appropriateness to the organization.

ISO 9004 Lists what top management should consider in establishing the quality policy and spells out conditions that will allow it to be used for QMS improvement.

ISO 9000 Defines a "quality policy" in 3.2.4 as a formal expression of the "overall intentions and direction of an organization related to quality . . . ." 2.5, Quality Policy and Quality Objectives, explains the policy's purpose and role and its relationship with the objectives.

Document Requirements:

Required:

Quality policy statement

Remember:

The quality policy will now need to include commitment to continuous improvement and to measurable objectives.

Internal Audit Questions:

- Is the quality policy documented?
- Does the quality policy provide a framework for establishing and reviewing the quality objectives?
- Does the quality policy demonstrate a commitment to meeting requirements and include provisions for continual improvement?
- Does management review the quality policy regularly for continuing suitability? Do the management review records indicate that the quality policy has been reviewed regularly for suitability?
- Do the quality objectives provide measurable evidence that the organization is achieving their stated quality policy?
- Are the members of the organization clear that the personnel know what the policy means and their role in carrying out the policy?

Management Summary:

Top management is responsible for setting this overarching vision for the QMS, from which the objectives will cascade, and must utilize the quality policy as a means of leading the organization toward improvement of its performance.

The policy must be ambitious but realistic and relevant to the organization.

Process Control Tip:

The best quality policies state not only how the company will focus its operations, but how the organization will measure those objectives to assure quality.

5.3 Quality Policy

Top management
must establish *and* maintain
a meaningful quality policy

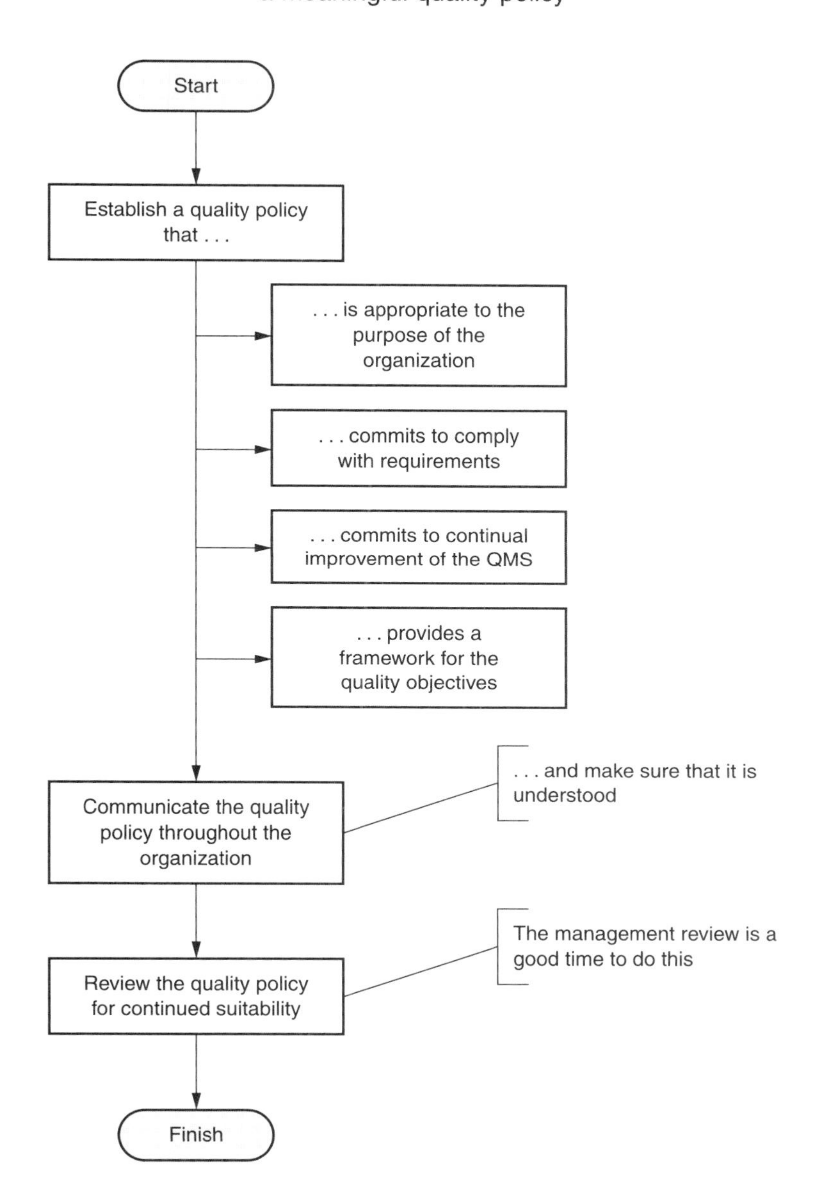

5.0 Management Responsibility

5.4 Planning

5.4.1 Quality Objectives

AS9100 Consistent, measurable quality objectives must be set under top management's auspices wherever they are required within the organization to satisfy product requirements, and based on the foundation provided in the quality policy.

ISO 9004 Indicates that quality objectives lead to improved organizational performance and recommends that top management establish them based on the organization's strategic planning and quality policy. Suggests factors worth considering in setting objectives and then communicating the objectives so everyone contributes to their achievement.

ISO 9000 Defines "quality objective" in 3.2.5 and explains in 2.5, Quality Policy and Quality Objectives, their purpose/role with the policy and the benefits of achieving objectives.

Document Requirements:

Required:

Quality policy objectives

Remember:

The need for objectives to be quantifiable in form as actual input on QMS effectiveness, which was only implied in ISO 9001:1994, is now a specific requirement.

Internal Audit Questions:

- Have measurable quality objectives been established at relevant functions and levels within the organization? How are these documented?
- Do quality objectives include meeting requirements for the organization's products and/or services?
- Has the organization identified the activities and processes required to meet the quality objectives (quality management system, product and/or service realization, verification processes)?

Management Summary:

The organization should use its strategic planning process and quality policy as a framework for setting quality objectives.

Top management sets quality objectives as a method for leading continual improvement within the organization.

Quality objectives must be measurable and communicated in such a way that all personnel can contribute to their achievement.

Quality objectives must be regularly and systematically reviewed and revised.

Process Control Tip:

Ensure that the elements of the quality policy exist in the hoshin plan.

5.4 Planning

5.4.1 Quality Objectives

Top management
must ensure that meaningful
quality objectives
are established

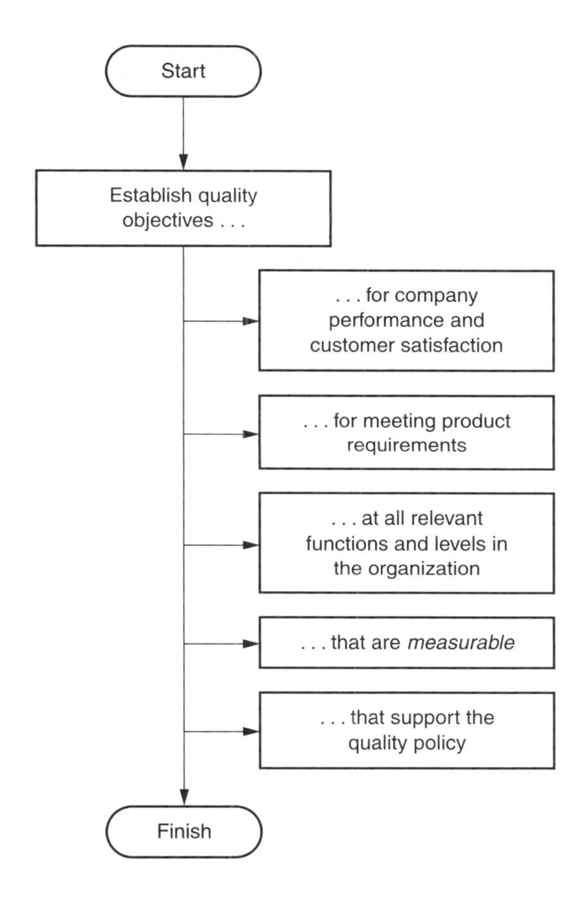

5.0 Management Responsibility

5.4 Planning

5.4.2 Quality Management System Planning

AS9100 Title is clarified from "Quality Planning" in the standard. Top management must ensure that planning relative to the QMS is conducted so as to conform with the requirements of 4.1, Quality Management System—General Requirements, and achieve the organization's quality objectives. This includes ensuring that QMS revisions do not compromise ISO 9001 conformance and effectiveness of the QMS.

ISO 9004 Discusses the broader concept of "quality planning" and the inputs and outputs that could make the QMS and quality planning effective and efficient in meeting an organization's quality objectives and strategic requirements.

ISO 9000 Defines "quality planning" in 3.2.9 as management activities "focused on setting quality objectives and specifying necessary operational processes and related resources to fulfill the quality objectives."

Document Requirements:

Remember:

- A documented plan is required that identifies the resources needed to achieve quality objectives.
- Changes are to be introduced without failures of the quality management system.

Internal Audit Questions:

- Has the organization identified and planned processes and resources for the quality system?
- Do the processes provide for the control of changes to the quality management system?
- As organizational changes occur, does quality planning take into account the needs of the organization?
- What objective evidence exists to demonstrate that quality planning contributes to continual improvement (for example, program briefs, action item in a management review record, or reference to an existing procedure)?

Management Summary:

Top management must establish quality plans for the organization.

Top management should plan processes effectively and efficiently to fulfill the organization's quality objectives.

Process Control Tip:

Original versions of ISO 9001 and AS9100 were organized by requirement elements, whereas the latest versions address requirements based on business processes. The intent was to support organizations by simplifying the relationship between QMS requirements and processes for better process control.

5.4 Planning

5.4.2 QMS Planning

Top management
is responsible for both the
planning and the integrity of the
quality management system

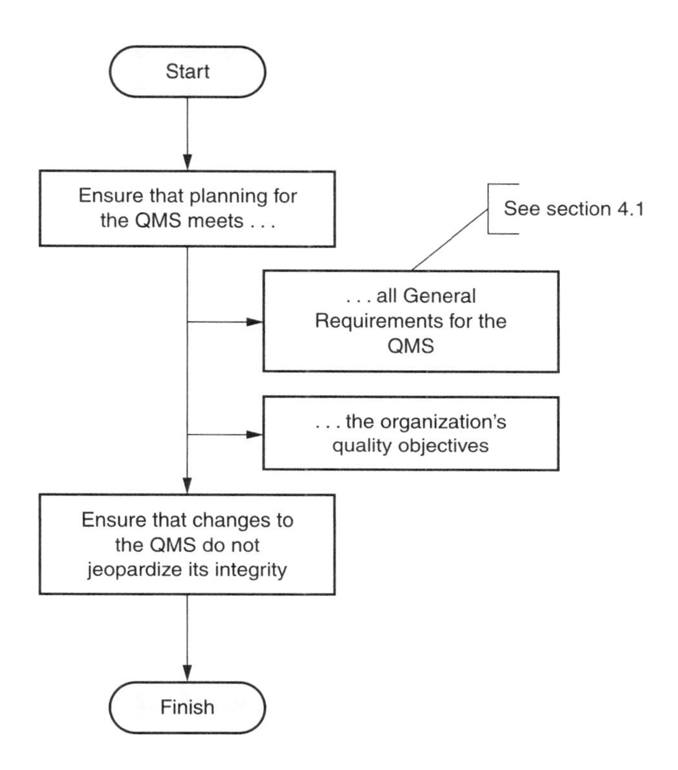

5.0 Management Responsibility

5.5 Responsibility, Authority, and Communication

5.5.1 Responsibility and Authority

AS9100 The responsibilities and authorities for management of an organization's QMS and the interrelationships are to be "defined and communicated" throughout the organization, with top management responsible for guaranteeing this is accomplished.

ISO 9004 Recommends that top management should "define and communicate" the responsibilities and authorities, with a range of employees empowered to gain their contributions to quality objective fulfillment and their buy-in to the QMS.

Document Requirements:

Remember:

- Organizational chart
- Job descriptions
- Other documents defining the authority and responsibility of each job function.

Internal Audit Questions:

- Are responsibilities and authorities defined and communicated to facilitate effective quality management?
- Are the quality objectives consistent with the stated quality policy and do they provide measurable evidence that the organization is achieving that policy?
- Are responsibilities and authorities defined and communicated to facilitate effective quality management?

Management Summary:

The processes within the organization must enable people to contribute to the achievement of the quality objectives, thus better aligning the business system with the management system.

Not all quality objectives will be suitable to and attainable by all operations or personnel, so top management must provide the direction and processes by which objectives can be set for each area and group of personnel.

Process Control Tip:

Use hoshin methodology to cascade top-level objectives down into the organization, with employees being accountable for measures applicable to their job description. Personnel like to be aware of what is planned for in the company and how their efforts roll up to the major business objectives and the bottom line.

Following this method also eases linkage of project results to strategy.

5.5 Responsibility, Authority, and Communication

5.5.1 Responsibility and Authority

Top management must see that responsibilities and authorities are defined and are communicated throughout the organization.

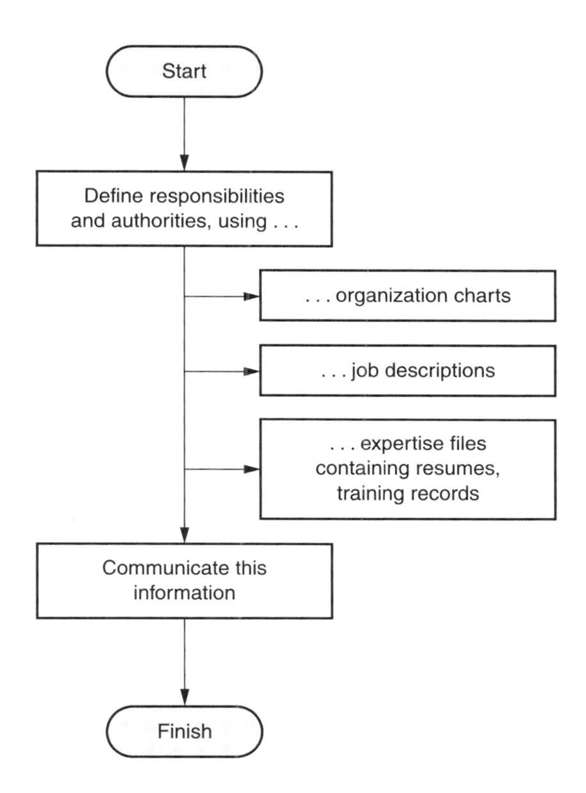

5.0 Management Responsibility

5.5 Responsibility, Authority, and Communication

5.5.2 Management Representative

AS9100 Top management must appoint a management-level representative who is to: ensure a QMS is put in place and kept in operation; report to top management on how well the QMS is functioning and actions required to correct/improve the QMS; and ensure customer requirements are communicated and understood by all affected employees (that is, listening to the voice of the customer). ***Organizationally, the representative must have the freedom to act on matters pertaining to quality.*** The representative can serve as a liaison to external parties on QMS issues.

ISO 9004 Advocates having the representative play the most active role in managing and overseeing the QMS and communicating with customers and others about the QMS.

Document Requirements:

Remember:

- No documents/documentation required, although a documented QMS planning process and/or records of planning activities are recommended
- Though not for creation, implementation, and revision of the QMS, a documented process for QMS planning will assist those involved in those planning activities and will help the organization, making a documented procedure and records of the QMS planning process useful.

Internal Audit Questions:

- Has the organization appointed one or more management representatives? Are the responsibilities and authorities of the management representative defined?
- As organizational changes occur, does QMS planning take into account the needs of the organization?
- Has the organization appointed one or more management representatives? Are the responsibilities and authorities of the management representative defined?
- Has the management representative implemented a system/program for communicating customer requirements to the organization?
- Is there objective evidence to demonstrate that this communication has taken place?
- Does the management representative report to top management on the performance of the quality management system?
- Does the management representative promote awareness of customer requirements throughout the organization?

Management Summary:

The role of the management representative is to enhance effective and efficient operation and improvement of the QMS.

The management representative has additional responsibility to promote awareness of customer requirements.

Ensure that planning takes place to produce and update an ISO 9001:2008–conforming QMS that meets the needs of the organization.

It is imperative that the representative have autonomy to act on matters of quality—their position should not be compromised by additional responsibilities that may conflict with achieving best quality.

Process Control Tip:

Ensure that your organization it set up to support accountability of the overall value stream. Measure management across departments, where applicable, to assure overall system effectiveness.

5.5 Responsibility, Authority, and Communication

5.5.2 Management Representative

Top management
must appoint a manager
authorized to oversee the
quality management system

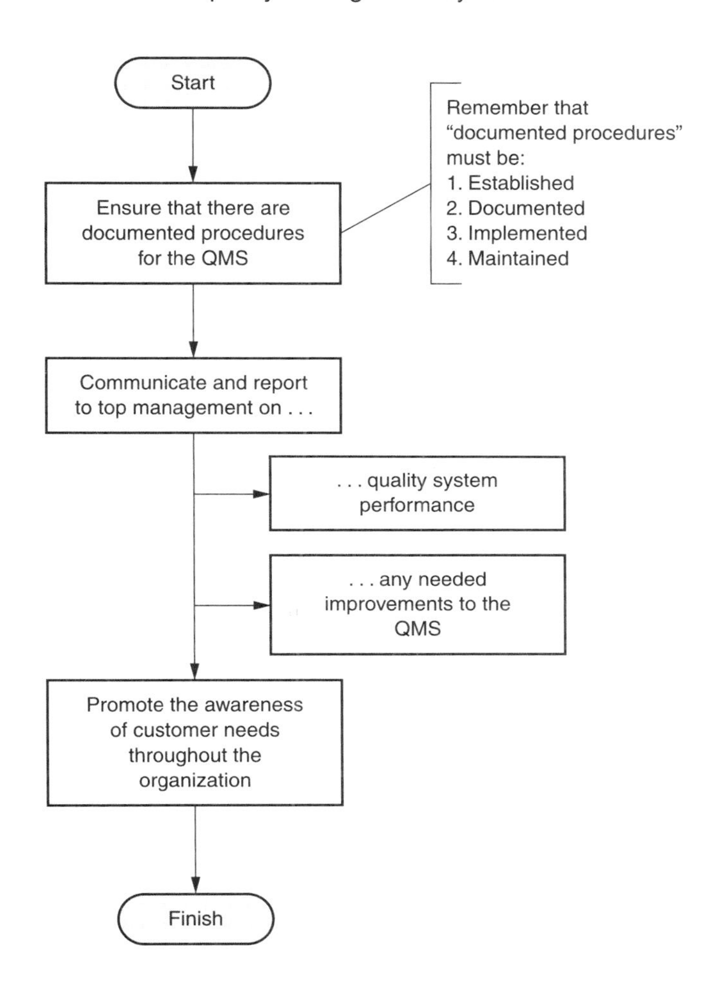

5.0 Management Responsibility

5.5 Responsibility, Authority, and Communication

5.5.3 Internal Communication

AS9100 Top management must ensure that the organization adopts communication processes that inform all within the organization about how effective the QMS is.

ISO 9004 Recommends management take direct responsibility for installing communication processes and including information on certain QMS elements and accomplishments to encourage employee participation in performance improvements, achievement of quality objectives, and providing QMS feedback. Lists ways to communicate with employees.

Document Requirements:

Remember:

No documents/documentation is specified, although documentation is the most effective means of defining and communicating QMS responsibilities and authorities. The following types of documentation can be used to define and communicate: documented procedures and records.

Internal Audit Questions:

- A functioning QMS within the organization will have defined responsibilities and authorities effectively known to, understood, and used by personnel throughout the organization. What objective evidence exists to document internal communication (document distribution lists and records, training records, minutes of management reviews, and so on)?
- Does the organization hold discussions with all employees to communicate processes of the quality management system and their effectiveness?

Management Summary:

- It is top management's responsibility to provide the processes and directives so that QMS responsibilities and authorities can be delegated to the appropriate personnel within the organization (where management does not actually determine those roles) and communicated so that all personnel understand who to contact for approvals and information and to whom to give feedback.
- Management must communicate the quality policy, requirements, objectives, and accomplishments to the organization.
- Management should actively encourage feedback and communication from people in the organization as a means of involvement.
- It is a new requirement for management to define and implement the communication process.

5.5 Responsibility, Authority, and Communication

5.5.3 Internal Communication

Top management must . . .

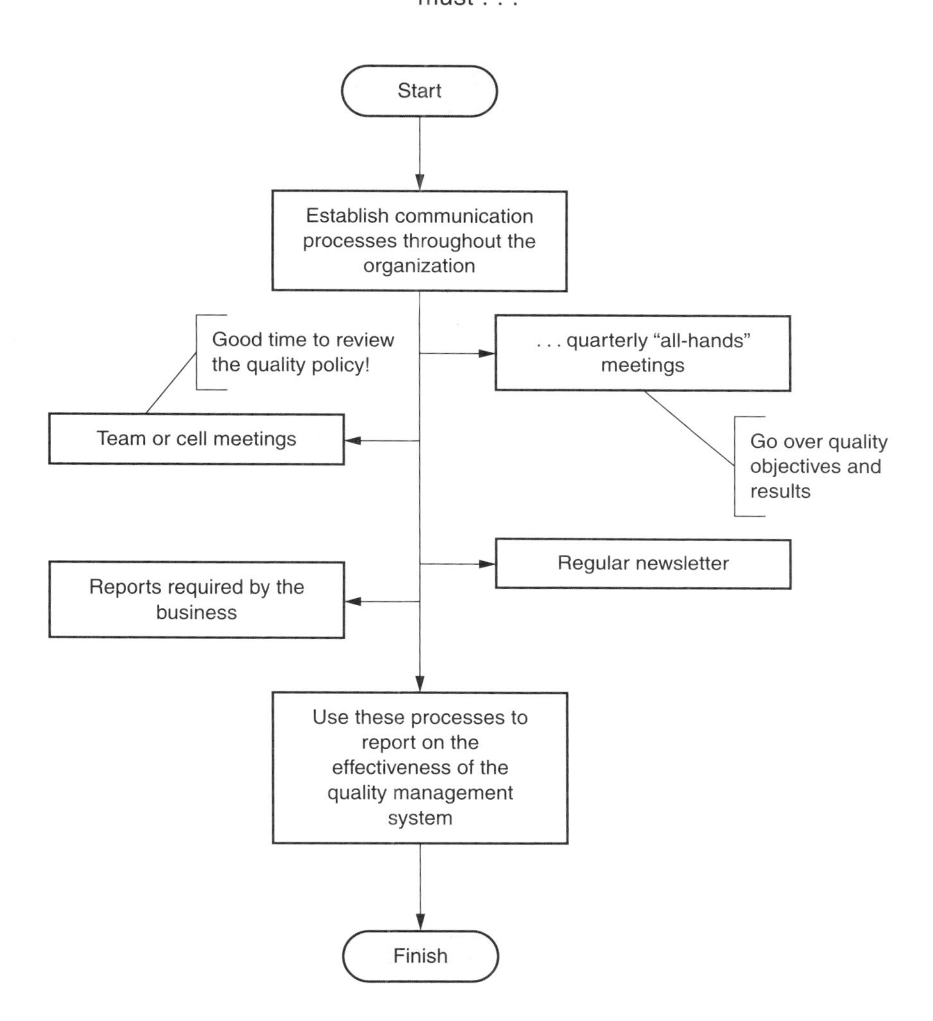

5.0 Management Responsibility

5.6 Management Review

5.6.1 General

AS9100 Top management is to review the QMS "at planned intervals" to verify that it is appropriate to the organization's needs, satisfies ISO 9001 requirements, and effectively meets customer requirements and employee needs. Meeting minutes of each review must be retained as quality records. During these reviews, management must evaluate potential QMS improvements and changes in response to nonconformities as well as achievements.

ISO 9004 Provides guidance on expanding management review to assess QMS efficiency and include review of all activities within the organization. The quality management principles should serve as the basis for "systematic control," with the planning of performance improvements for the whole organization as an end result.

ISO 9000 In 2.8.3, Reviewing the Quality Management System, explores what top management's role is intended to be in the reviews.

Document Requirements:

Required:

Records of reviews

Remember:

Management may wish to hold more frequent reviews to address additional items and to demonstrate commitment to continual improvements.

Internal Audit Questions:

- Has the management representative overseen creation, implementation, and maintenance of the QMS processes required by the standard?
- Has top management responded appropriately to defined reports? Has the management representative implemented a system/program for communicating customer requirements to the organization that ensures awareness?
- Does management review the quality management system at planned intervals to ensure its continuing suitability, adequacy, and effectiveness?
- Do management reviews include evaluation of the need for changes to the organization's quality management system, quality policy, and quality objectives?
- Are results of management reviews recorded and maintained as records?

Management Summary:

Top management must select one of its members based on the ability to fulfill the obligations of the management representative.

Top management must work with the management representative to establish a regularly scheduled report from the representative and must be committed to act in response to those reports.

Top management should develop an expanded management review process that extends to the entire organization.

Management review should be a platform to exchange new ideas with open discussion.

The frequency of management reviews should be determined by the needs of the organization.

Management review inputs should result in changes to the process that will increase the existing efficiency and effectiveness of the QMS.

5.6 Management Review

5.6.1 General

Top management must . . .

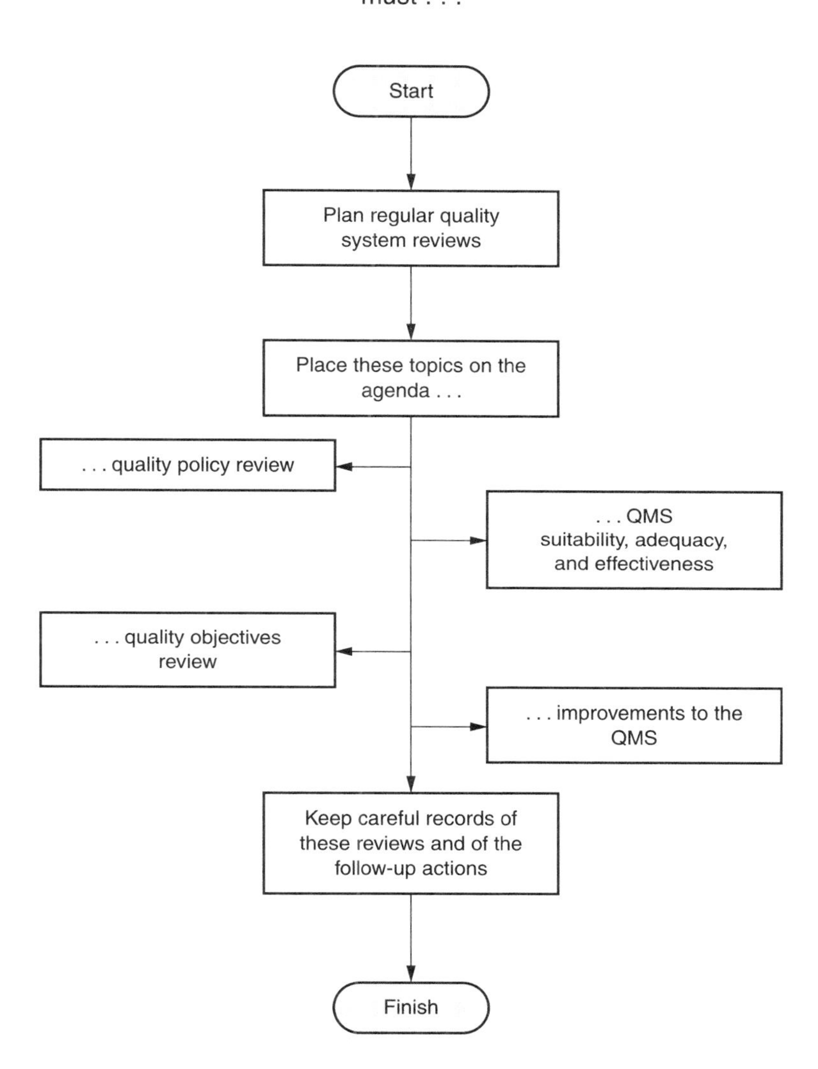

5.0 Management Responsibility

5.6 Management Review

5.6.2 Review Input

AS9100 Management must include seven categories of data as mandatory review inputs:

a) Results of audits
b) Customer feedback
c) Process performance and product conformity
d) Status of preventive and corrective actions
e) Follow-up actions from previous management reviews
f) Changes that could affect the quality management system
g) Recommendations for improvement

ISO 9004 Includes as suggested inputs 13 categories of data, ranging from QMS results beyond the ISO 9001 requirements to non-quality business considerations.

Document Requirements:

Remember:

There are specified document requirements. The organization has now clearly defined items that must establish communications approaches that would best be documented and where included on the actual communications must take place through some form of documentation (for example, QMS effectiveness memos, reports, and/or intranet or Web-based communications). Records of these communications should be kept.

Internal Audit Questions:

- Does management review input include: audit results, customer feedback, process performance and product performance, preventive and corrective action status, follow-up actions from earlier management reviews, and changes that could affect the quality management system?
- Is top management involved within the QMS structure?

Management Summary:

Management review inputs should include the status of relevant internal and external factors, as well as results of existing processes.

Inform the organization of the quality policy objectives and resulting processes to be followed. Top management should seek verification that QMS effectiveness—and the need for improvements—are known and understood throughout the organization.

To foster employees in the maintenance and improvement of the QMS top management must ensure that communication processes are put into use.

Process Control Tip:

Many organizations perform management review as a separate quality meeting instead of infusing it into the overall strategic and operational meetings necessary for proper business functioning throughout the year.

Utilization of hoshin principles can help infuse AS9100 requirements into the major business objectives.

5.6 Management Review

5.6.2 Review Input

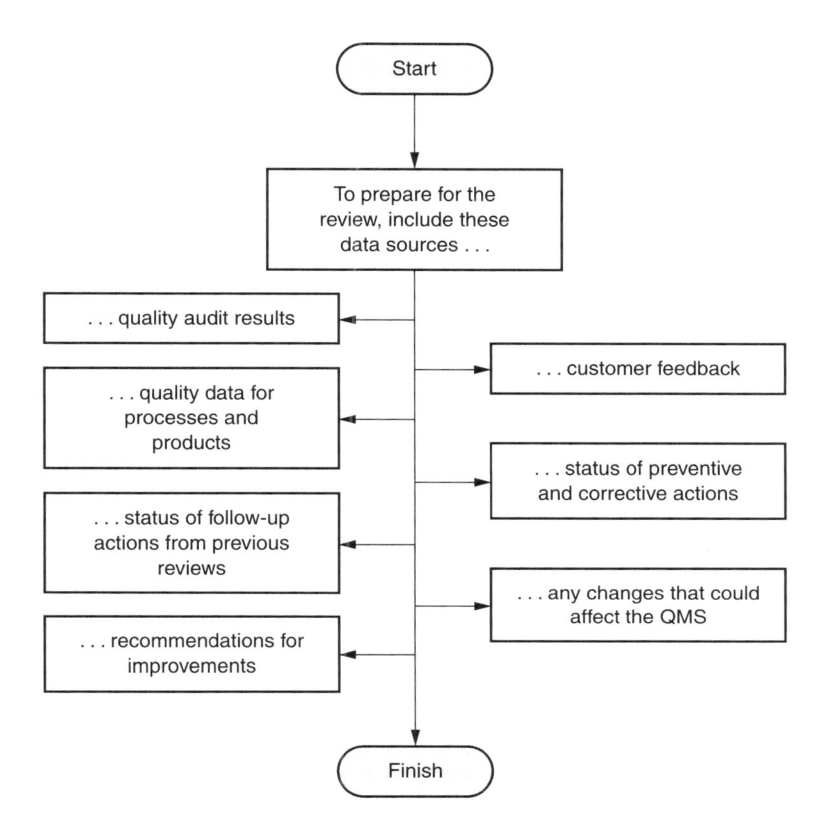

5.0 Management Responsibility

5.6 Management Review

5.6.3 Review Output

AS9100	The output from management reviews must include any "decisions and actions" in three categories relating to the QMS and customer satisfaction, including product improvements linked to customer specifications.
ISO 9004	Provides guidance on how top management can use outputs from an "expanded management review" to identify potential performance improvements and to motivate employees by establishing new objectives. Suggests six categories of outputs linked to QMS efficiency that could be obtained.

Document Requirements:

Required:

Management review meeting minutes will be required to address these improvement activities and the resources required to initiate these process actions.

Internal Audit Questions:	*Management Summary:*
• Do management review outputs include actions related to improving the quality management system and its processes? • Do management review outputs include actions related to improvement of product related to customer requirements?	The strategic planning process should determine the frequency of management reviews. Management reviews should result in changes to the organization's processes so as to increase the existing efficiency and effectiveness of the QMS. Adequate records should be kept to provide traceability and facilitate the evaluation process in future management reviews.

5.6 Management Review

5.6.3 Review Output

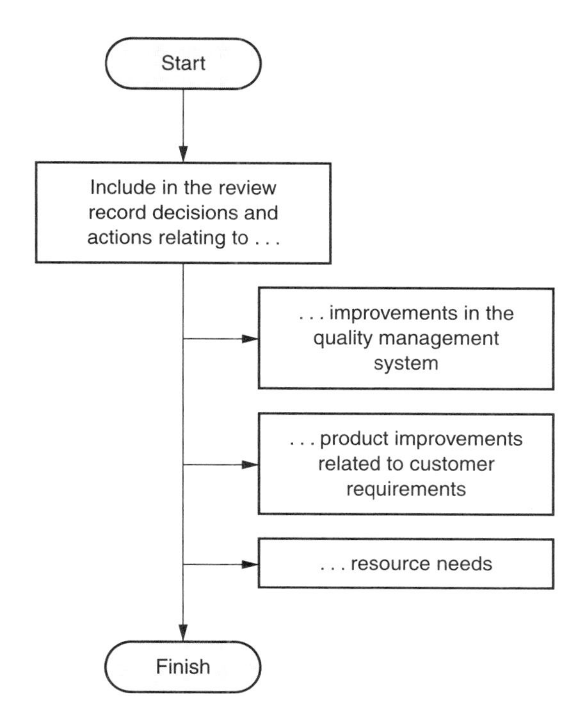

REVISIONS TO AS9100:2009 (REVISION C), CLAUSE 5

(Source: Larry Whittington)

5. Management Responsibility

5.1 Management Commitment

No changes in AS9100C clause 5.1.

5.2 Customer Focus

AS9100C added this requirement to measure product conformity and on-time delivery performance. It establishes a clear link between the quality management system and organizational performance:

> *Top management shall ensure that product conformity and on-time delivery performance are measured and that appropriate action is taken if planned results are not, or will not be, achieved.*

5.3 Quality Policy

5.4 Planning

No changes in AS9100C clauses 5.3 or 5.4.

5.5 Responsibility, Authority, and Communication

5.5.1 Responsibility and Authority

No changes in AS9100C clause 5.5.1.

5.5.2 Management Representative

Most organizations already appoint a management representative that is a member of their own management team. The following change clarifies that requirement:

> *Top management shall appoint a member of the organization's management who, irrespective of other responsibilities, shall have responsibility and authority that includes:*

Some companies in the past have outsourced the management representative role to someone in a different organization, or even to their consultant. This text change may be aimed at that practice.

AS9100C revised subclause (d) to ensure that the appointed management representative has sufficient clout in the organization to be listened to by top management:

> *d) the organizational freedom and unrestricted access to top management to resolve ~~matters pertaining to~~ quality management issues.*

5.5.3 Internal Communication

No changes in AS9100C clause 5.5.3.

5.6 Management Review

No changes in AS9100C clause 5.6.

SECTION 6: RESOURCE MANAGEMENT

6.1—Provision of Resources
6.2—Human Resources
6.3—Infrastructure
6.4—Work Environment

6.0 Resource Management

6.1 Provision of Resources

AS9100 The organization is to identify and make available the means for creation, continued operation, and ongoing increases in the effectiveness of the QMS and improvement in the level of customer satisfaction.

ISO 9004 Defines possible resources to be provided and recommends that top management provide the resources to pursue strategic goals and achieve objectives. Includes a list of possible resources and considerations to help improve organizational performance.

Document Requirements:

Remember:

Management will be required to formalize its resource planning process to ensure timely availability.

Internal Audit Questions:

- Has the organization determined and provided the resources necessary to implement the processes of the quality management system?
- Has the organization determined and provided the resources necessary to improve the processes of the quality management system?
- Has the organization determined and provided the resources necessary to address customer satisfaction?
- Are the resources provided in a timely manner?
  - Internal and other QMS audit reports
  - Data derived from customer feedback processes
  - Process and product performance monitoring and measurement results
  - Reports on preventive and corrective actions undertaken and their status
  - The status of actions undertaken in response to earlier management reviews
  - Changes that could affect the quality management system

Management Summary:

Top management should provide adequate resources to implement the QMS and achieve its objectives.

Top management must direct those having responsibility and/or authority for the QMS and related organizational operations to provide specified inputs to be used in management reviews of the QMS, including the frequency of such inputs and the processes and forms in which the inputs are to be provided.

Top management should consider the number of inputs to be reviewed in allocating adequate time for the reviews to evaluate them (and any time prior to the meetings for participants to study the inputs).

Top management should approach evaluation of the inputs with consideration of what the quality policy and quality objectives state and what the organization's business performance measurements indicate.

Resources may include:

- People
- Infrastructure
- Work environment
- Information
- Suppliers and partners
- Energy and financial resources

6.1 Provision of Resources

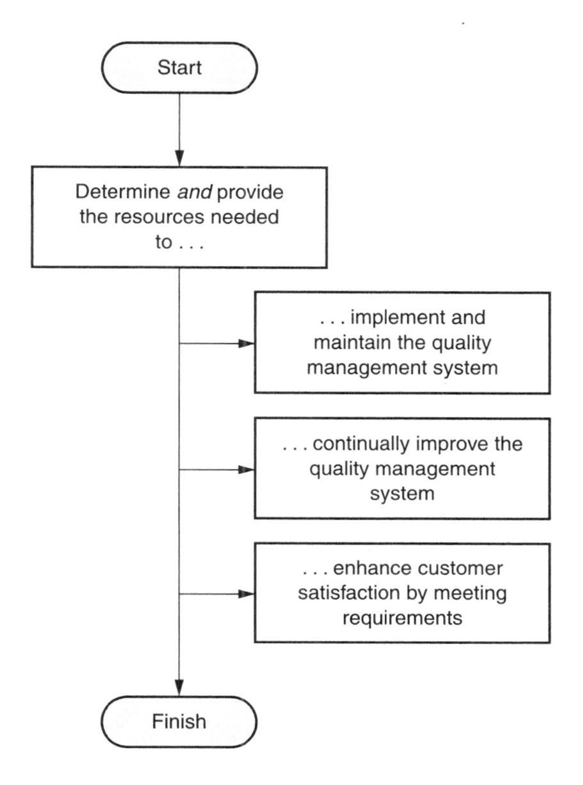

6.0 Resource Management

6.2 Human Resources

6.2.1 General

AS9100 Personnel whose actions impact product quality must be competent to work on products based on "education, training, skills, and experience."

ISO 9004 For an organization to achieve performance improvements, ISO 9004 recommends 12 types of activities to pursue to "encourage the involvement and development of its people."

Document Requirements:

Remember:

- Records that are created by the activities to assure competency may need to be controlled per clause 4.2.4, Control of Records.
- Records of the management reviews. While the need for documentation of any decisions reached and actions determined as outputs of the reviews is not specified, some form of documentation will be necessary in most cases to communicate the improvements that top management wants to the QMS and/or product.

Internal Audit Questions:

- Do personnel with assigned responsibilities within the quality management system have competency in the appropriate education, training, skills, and experience?
- Have decisions on how to better meet and enhance customer satisfaction been made?
- Has top management made changes to the quality policy and/or quality objectives as a result of management reviews?

Management Summary:

Improvements to the effectiveness and efficiency of the organization should be accomplished through the involvement and support of the people within the organization.

Top management must use management reviews as a forum in which to ensure that the QMS remains effective in meeting the needs of the organization and its customers and to determine what continual improvement efforts will be undertaken before the next review.

It is expected that top management will reach decisions about improvements and direct the organization to take steps to achieve those improvements as outputs of management reviews.

6.2 Human Resources

6.2.1 General

Anyone who can affect product quality must be competent to do so

6.0 Resource Management

6.2 Human Resources

6.2.2 Competence, Training, and Awareness

AS9100 The organization must identify qualifications required for each position affecting product quality, take appropriate actions to ensure personnel competency in those positions, assess the effectiveness of these efforts, and make personnel aware of how they affect quality and what they can do to achieve quality objectives. The organization must document personnel competence and actions taken relative to competence and retain these documents as quality records.

ISO 9004 Recommends that an organization analyze present and future needs and determine gaps between needs and existing personnel capabilities, with consideration of five sources for determining competency needs. Explores what to consider in terms of education and training relative to meeting quality objectives and raising awareness. Defines possible subjects to include in education and training and to be addressed in training plans.

Document Requirements:

Required:

Record (e)

Internal Audit Questions:

- Does the organization identify the competency needs of personnel performing activities affecting quality (including additional training needs)?
- Does the organization provide training to satisfy these needs?
- Does the organization evaluate the effectiveness of the training?
- Does the organization ensure that employees are aware of the importance of their activities and how they contribute to the achievement of the quality objectives?
- Does the organization maintain records of education, experience, training, and qualifications (for example, banners, posters, bulletin boards, or intranet sites)?

Management Summary:

Management should analyze both the present and expected competence needs within the organization.

Management should plan for education and training needs and take into account the organization's processes, stages of development, and culture.

Process Control Tip:

Hoshin planning ("policy deployment") can be linked to the AS9100 standard to reduce duplication of effort. With this Resource Management section in mind, assure that your hoshin business fundamental objectives cascade down to the level of providing the accountability of each employee to their job description.

Personal growth opportunities come from hoshin breakthrough strategies, as those roles necessary to support breakthroughs planned for the year are identified, with skill gaps assessed and closed through necessary training. (Refer to Section 5.3.)

6.2 Human Resources

6.2.2 Competence, Training, and Awareness

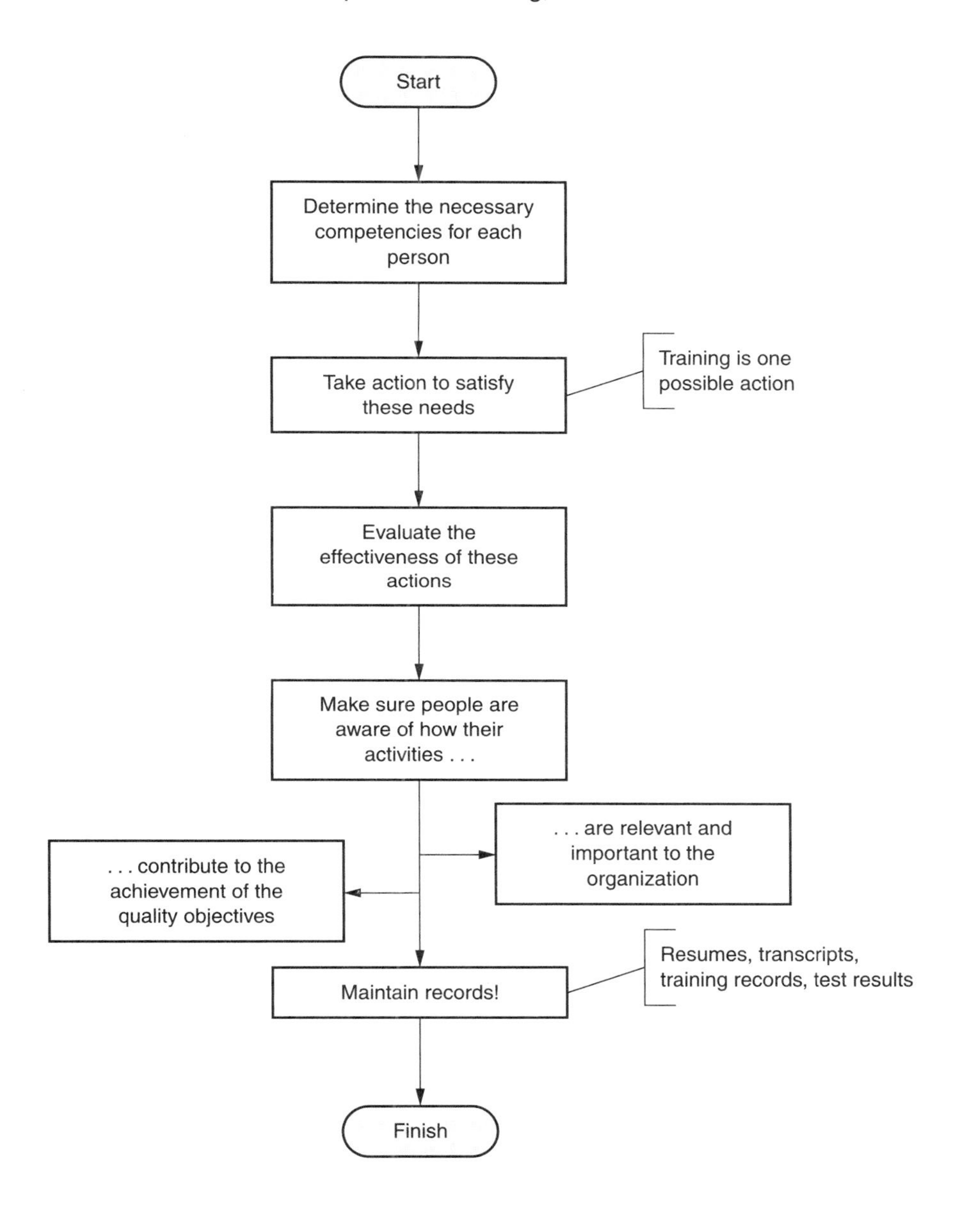

6.0 Resource Management

6.3 Infrastructure

AS9100 States that the organization is responsible for the establishment and upkeep of the "infrastructure" required to provide customers with products according to specifications, with examples of what is meant by infrastructure spelled out.

ISO 9004 Recommends four factors that management should address in defining the required infrastructure and advises an organization to consider natural phenomena and the risks they pose to its operations in developing and keeping up the infrastructure.

ISO 9000 Defines "infrastructure" in 3.3.3 as a "system of facilities, equipment, and services needed for the operation of an organization."

Document Requirements:

Remember:

The records that are created by the activities to fulfill the requirements of this clause may need to be controlled per clause 4.2.4, Control of Records.

Internal Audit Questions:

- Does the organization identify, provide, and maintain the work space and associated facilities needed to achieve conformity of product?
- Does the organization identify, provide, and maintain the supporting services, equipment, hardware, and software it needs to achieve conformity of product?

Management Summary:

Infrastructure includes plant, work space, tools, equipment, support services, information/ communication technology, and transport facilities.

Process Control Tip:

Utilize lean methodologies to identify what is truly needed in each area to add value, thereby providing the opportunity to remove non-value-added infrastructure. (Refer to Section 6.4.)

6.3 Infrastructure

6.4 Work Environment

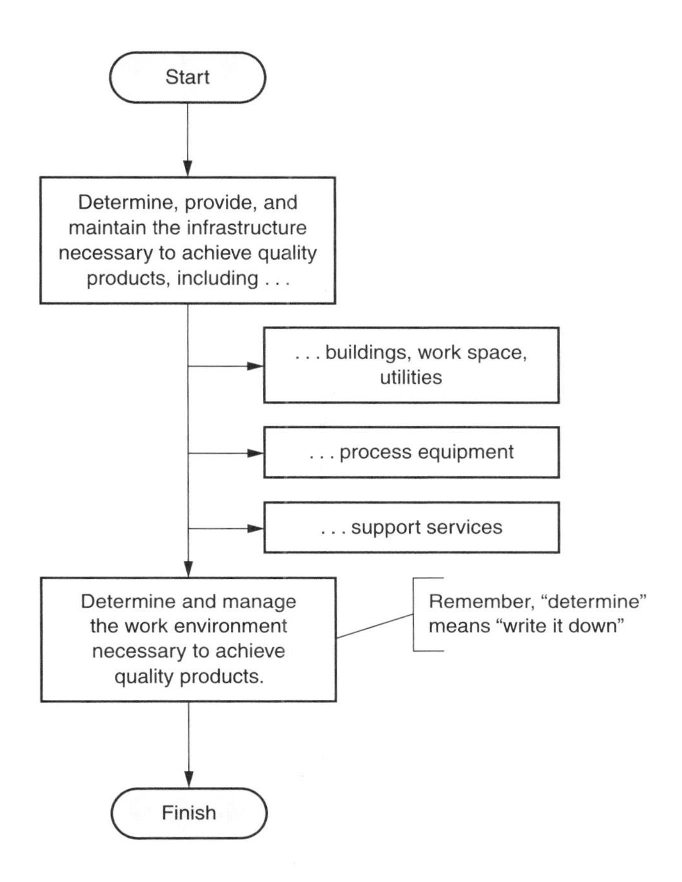

6.0 Resource Management

6.4 Work Environment

AS9100 The organization is required to maintain the work environment in a state that permits product to satisfy requirements.

ISO 9004 Encourages management to maintain the work environment so as to improve employee "motivation, satisfaction, and performance" and thereby improve organizational performance. Identifies seven aspects of the work environment that should be evaluated in developing/revising the work environment.

ISO 9000 Defines "work environment" in 3.3.4 as a "set of conditions under which work is performed," with a list of conditions to be included in the scope of the work environment. Clarifies this as the conditions under which work is performed, not conditions under which people perform work.

Document Requirements:

Remember:

The records that are created by the activities to fulfill the requirements of this clause may need to be controlled per clause 4.2.4, Control of Records.

Internal Audit Questions:

- Does the organization identify and manage the human and physical factors of the work environment needed to achieve conformity of product?

Management Summary:

Using a combination of human and physical factors, management should consider:

- Creative work methods and opportunities for greater involvement
- Ergonomics
- Heat, humidity, light, airflow
- Hygiene, cleanliness, noise, vibration, and pollution
- Electrostatic discharge

Process Control Tip:

The core premise of lean support is simplifying and optimizing work area functionality through 5S/visual/layout principles. (Refer to Section 6.3.)

Specification requirements such as electrostatic discharge should be included in control plans. (Refer to clause 4.2.4.)

6.3 Infrastructure

6.4 Work Environment

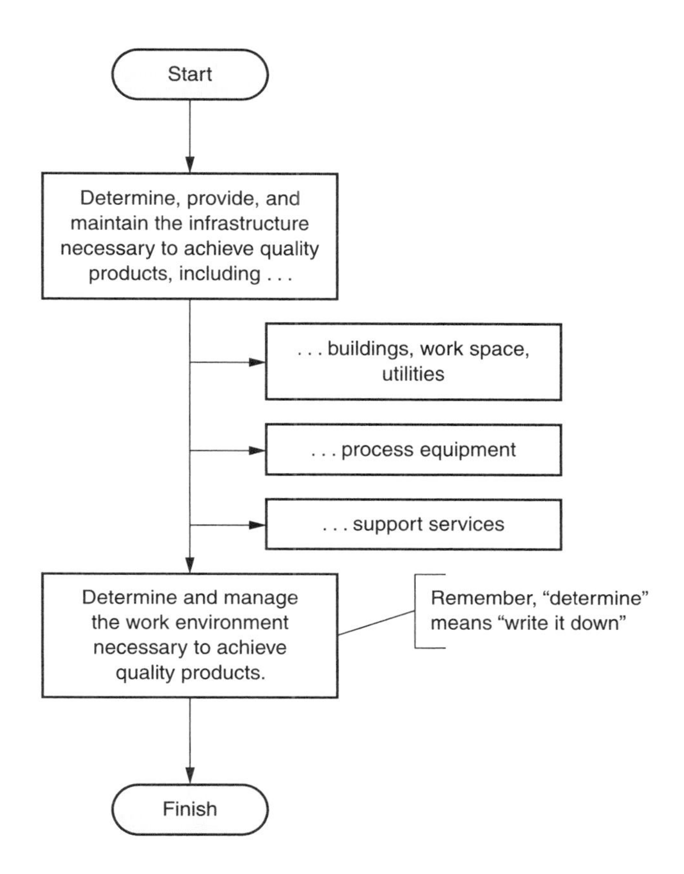

REVISIONS TO AS9100:2009 (REVISION C), CLAUSE 6

(Source: Larry Whittington)

6. Resource Management

6.1 Provision of Resources

No changes in AS9100C clause 6.1.

6.2 Human Resources

6.2.1 General

The revision below changes from work affecting "product quality" to work affecting "conformity to product requirements." Since quality is the degree to which a set of inherent characteristics fulfills requirements, then product quality would be the degree of conformity to product requirements.

> *Personnel performing work affecting conformity to product ~~quality~~ requirements shall be competent on the basis of appropriate education, training, skills and experience.*

The revision above should not be viewed as a new requirement. Anyone performing, verifying, or managing work within the scope of the quality management system, including supporting services, can affect conformity to product requirements. A new Note has been added to 6.2.1 to explain this point:

> *NOTE: Conformity to product requirements can be affected directly or indirectly by personnel performing any task within the quality management system.*

6.2.2 Competence, Training, and Awareness~~, and Training~~

This clause title has been changed from "Competence, Awareness, and Training" to "Competence, Training, and Awareness." Awareness comes from some form of training and should be last in the title. And that is also the sequence of the requirements as listed within clause 6.2.2.

The change in 6.2.1 from "product quality" to "product requirements" has been made to this subclause:

> *a) determine the necessary competence for personnel performing work affecting conformity to product ~~quality~~ requirements,*

Use below of the phrase "where applicable" recognizes that training or other actions may not be necessary, since individuals may already have the necessary competence. And, since "these needs" could be taken out of context, the requirement has been revised to specifically mention competence:

> *b) where applicable, provide training or take other actions to ~~satisfy these needs~~ achieve the necessary competence,*

6.3 Infrastructure

The only change in 6.3 was to add "information systems" as an example of a supporting service:

c) supporting services (such as transport, ~~or~~ communication, or information systems).

6.4 Work Environment

AS9100B included a Note that listed factors that may affect product conformity. When ISO 9001:2008 added the following Note (included in AS9100C) there was no longer a need for a separate AS9100 Note.

NOTE: The term "work environment" relates to those conditions under which work is performed including physical, environmental, and other factors (such as noise, temperature, humidity, lighting, or weather).

~~NOTE: Factors that may affect the conformity of the product include temperature, humidity, lighting, cleanliness, protection from electrostatic discharge, etc.~~

SECTION 7: PRODUCT REALIZATION

7.1—Planning of Product Realization
7.2—Customer-Related Processes
7.3—Design and Development
7.4—Purchasing
7.5—Production and Service Provision
7.6—Control of Monitoring and Measuring Devices

7.0 Product Realization

7.1 Planning of Product Realization

AS9100 The organization is to engage in planning to create processes to achieve "product realization"—the processes required to design, develop, produce, deliver, and/or service a product—in conformance with other QMS requirements. Outcomes are to be identified as a result of the planning process, including: a product's quality objectives; processes, documentation, and resources for the product; procedures acceptable to the customer that will ensure that the product meets specifications; records that prove that the processes and product satisfy requirements. The outputs of the planning process must be appropriate for use by the organization based on how it functions, ***and adequate resources are to be provided for operation and maintenance.***

ISO 9004 Recommends that top management take steps to ensure the effectiveness and efficiency of all product realization and support processes and examines these processes as both inputs and outputs, with consideration of the roles of documentation, personnel, and operating plans. Examples are provided of support processes, input issues, and topics to be covered in process performance reviews.

ISO 9000 Defines "inspection," "test," "verification," and "validation" in 3.8.2–3.8.5.

Document Requirements:

Required:

Record (d)

Remember:

A documented procedure may be needed for describing how the planning is accomplished.

Internal Audit Questions:	*Management Summary:*
• Have documented quality plans for product realization processes and product validation been established (drawings, specifications, materials, process flow diagrams, process flowcharts, setup sheets, validation reports, and so on)? • Is there evidence of planning of production processes? • Does the planning encompass all product realization processes? • Is the planning consistent with other elements of the quality management system? • Does documentation of product realization exist? • Are product realization resources and facilities defined during the planning process? Are they adequate? • Does the planning define the records that are prepared to show confidence in the conformity of the processes and final product?	For effective operation, management should staff appropriately and recognize that a process output might become the input to another process. Verification results and process validations should be considered as inputs to a process. Effective processes can only offer increased benefits, improved customer satisfaction and use of resources, and waste reduction. Process inputs, which can be internal or external to the organization, include: • Competence of people • Documentation • Equipment capability and monitoring • Health, safety, and work environment. ***Process Control Tip:*** ***1. Effective control plans have Six Sigma process maps as the basis for identifying the key inputs to control for each process step.*** ***2. Lean methods help reduce non-value-added work, thereby increasing productivity and in turn alleviating inadequate staffing issues.***

7.1 Planning of Product Realization

The organization must define its own processes for product realization.

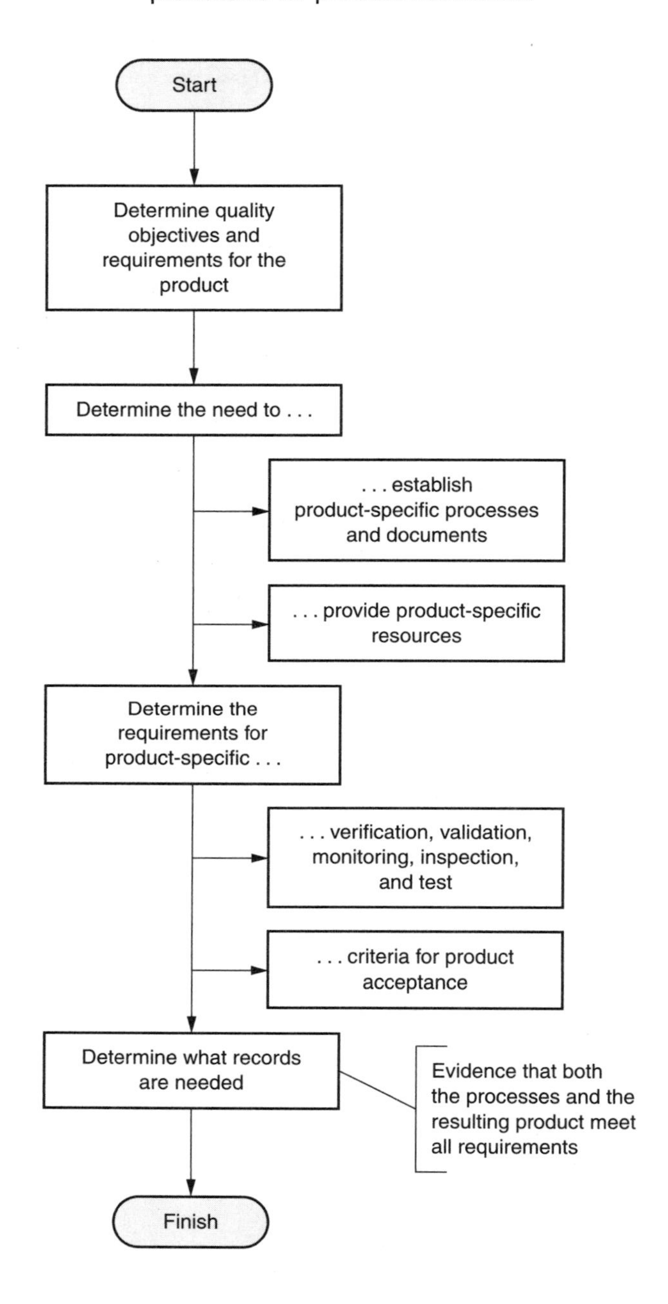

7.0 Product Realization:

7.2 Customer-Related Processes

7.2.1 Determination of Requirements Related to the Product

AS9100 The organization must identify all requirements of a product, including those from customer specifications, unspecified but necessary for the product to work properly, imposed by legal and regulatory authorities, and deemed necessary by the organization.

ISO 9004 Recommends that management consider the needs and expectations of all interested parties, including the customer, in determining other requirements for a product. (Note: This applies for 7.2.2 and 7.2.3 below.)

ISO 9000 Defines "requirement" in 3.1.2 as a "need or expectation that is stated, generally implied, or obligatory," with four Notes clarifying the definition.

Document Requirements:

Remember:

The organization should consider having a documented procedure to determine customer requirements. They are a vital part in achieving customer specifications, needs, and expectations.

Internal Audit Questions:

- Has the organization determined customer requirements and regulatory requirements, and so on?
- Have processes been established to determine both specified and unspecified customer requirements?
- Do records exist to provide evidence that customer requirements have been determined (for example, contract review records, service records, customer feedback records, surveys, market testing reports, design input and design validation records, identified legal requirements, and so on?

Management Summary:

Management should consider the following to ensure that the customer's needs and expectations are met:

- Operating conditions for the product
- Use or application of the product
- Disposal of the product
- Life cycle of the product
- Environmental impact of the product

Process Control Tip:

Consider Design for Six Sigma methodology—specifically quality function deployment (QFD)—to best ensure that customer specifications and regulatory and material/product requirements are embedded into the design from the start. (Refer to Sections 7.2.2, 7.3.1.)

7.2 Customer-Related Processes

7.2.1 Determination of Requirements Related to the Product

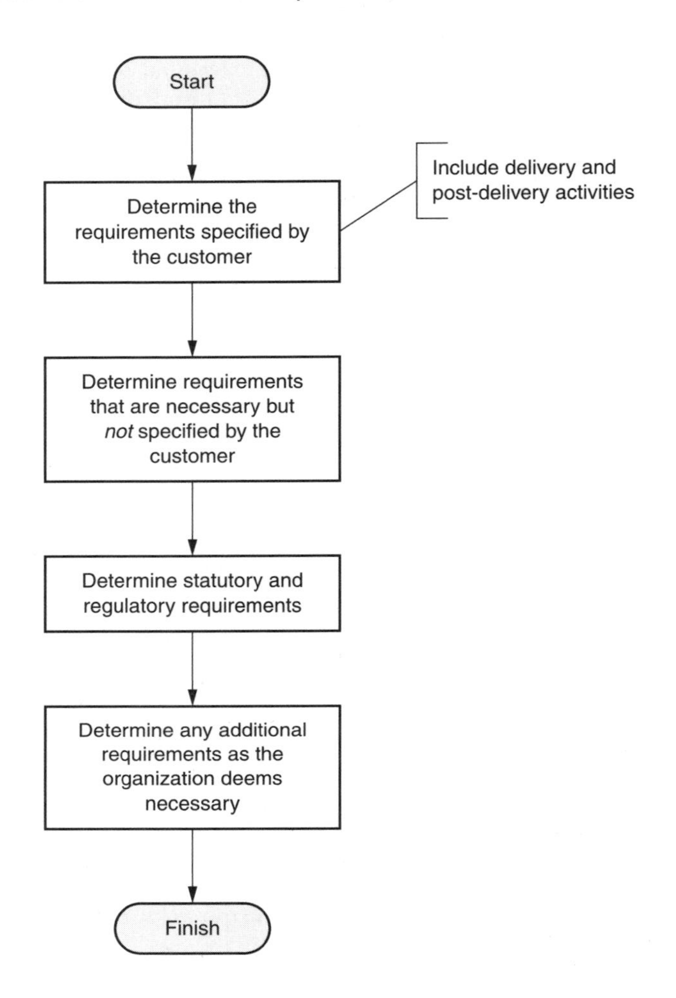

7.0 Product Realization

7.2 Customer-Related Processes

7.2.2 Review of Requirements Related to the Product

AS9100 Product requirements, whether involving a bid request, a new contract order, or an amendment to an existing contract, must be reviewed by the organization to ensure that the requirements and/or any changes are clearly understood and can be met, including in situations where the customer provides no written specifications. The review process results are to be treated as quality records, which must be updated whenever requirements change—with changes relayed to "relevant personnel" to ensure that customer requirements are met effectively. ***Evaluate what risks exist in complying with contractual and product requirements.***

ISO 9000 Defines "review" in 3.8.7 as "activity undertaken to determine the suitability, adequacy, and effectiveness of the subject matter to achieve established objectives."

Document Requirements:

Required:

Record

Remember:

Consider having a documented procedure for the review of customer requirements and how these requirements can be met.

Internal Audit Questions:

- Is there a process that requires the review of identified customer requirements before commitment to supply a product to the customer?
- Is there a process that requires the review of quotes and orders for adequacy of the definition of requirements?
- Is there a process for handling review of verbal orders?
- Is there a process to handle resolution of differences between quotations and orders?
- Is there a process for handling changes to product requirements?
- Are records of quote, tender, contract, and order review available and maintained?

Management Summary:

Product and process changes should be identified, recorded, evaluated, reviewed, and controlled in order to understand the effect on other processes and the needs and expectations of customers.

The organization should define the authority for initiating process changes in order to maintain control.

Process Control Tip:

Risk assessment utilizing Six Sigma [or failure mode and effects analysis (FMEA)] is key to identifying which process steps require better prevention or detection of failure, and is a core tool in the development of robust control plans.

Contract review should be designed to be the conduit linking contractual requirements to processes and designs—ensuring that changes are not made without knowledge of those requirements. (Refer to Sections 4.2, 7.3.1.)

Every control plan includes a change control plan. Contract review should own governance of requirement flow-down within the organization.

7.2 Customer-Related Processes

7.2.2 Review of Requirements Related to the Product

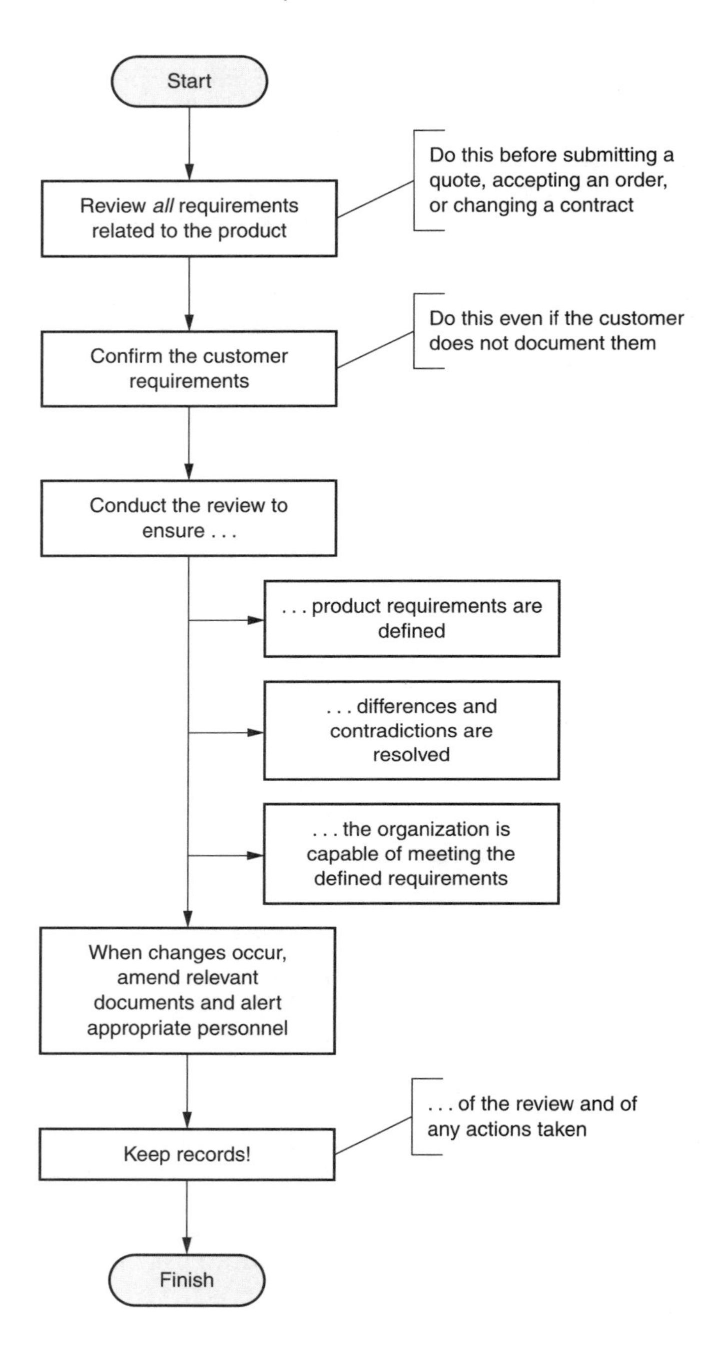

7.0 Product Realization

7.2 Customer-Related Processes

7.2.3 Customer Communication

AS9100 The organization must establish procedures for effective communication with customers regarding product specifications and the receipt and handling of product orders and amendments and any questions regarding these, and for obtaining customer feedback on products and related services.

Document Requirements:

Recommendation not required in Process Control Tip below.

Internal Audit Questions:	*Management Summary:*
• Is there a process in place to communicate with customers regarding product information, enquiries, contracts, order handling (including amendments), and customer feedback, including customer complaints?	The organization should implement and maintain a process to translate customer needs and expectations into requirements for the organization. The organization should fully understand the customer requirements prior to acceptance. ***Process Control Tip:*** ***Communication plans are essential to ensuring that appropriate individuals are accountable for their role in ensuring effective communication with the customer and back to the organization.***

7.2 Customer-Related Processes

7.2.3 Customer Communication

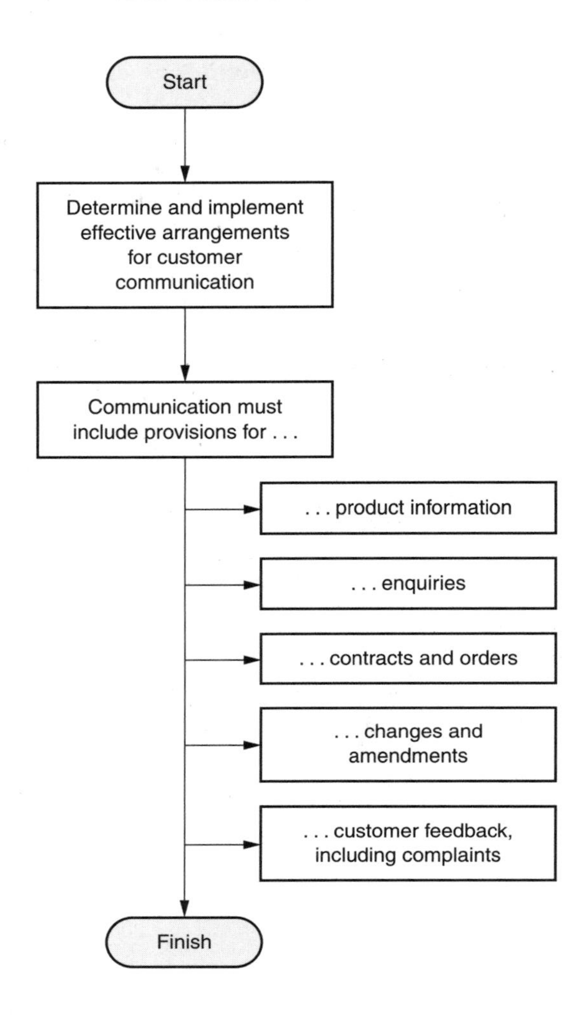

7.0 Product Realization

7.3 Design and Development

7.3.1 Design and Development Planning

AS9100 The processes involved in design and development of a product must be planned out and managed by the organization to ensure that each group involved with its design and development is engaged at all appropriate stages—and everyone involved understands who is responsible for which activities at what times—and that proper evaluations of prototypes, and so on, are conducted at appropriate stages—***with respect to the organization, task sequence, mandatory steps, significant stages, and configuration control.*** The output from this planning must be revised to reflect the product's progress through the design and development process.

Where appropriate, due to complexity, the organization shall give consideration to the following activities:

- ***Structuring the design effort into significant elements.***
- ***For each element, analyzing the tasks and the necessary resources for its design and development. This analysis shall consider an identified responsible person, design content, input data, planning constraints, and performance conditions. The input data specific to each element shall be reviewed to ensure consistency with requirements.***

The different design and development tasks to be carried out shall be defined according to specified safety or functional objectives of the product in accordance with customer and/or regulatory authority requirements.

ISO 9004 Suggests that top management should consider a number of aspects and techniques not directly related to customer specifications for the product in planning its design and development, including the design of the organization's processes. Recommends taking steps to "identify and mitigate potential risk" to product users and those affected by the organization's processes, providing a list of six risk assessment tools that could be applied in planning design and development.

ISO 9000 Defines "design and development" in 3.4.4 as a "set of processes that transforms requirements into specified characteristics or into the specification of a product, process, or system."

Document Requirements:

None

<table>
<tr><th>Internal Audit Questions:</th><th>Management Summary:</th></tr>
<tr><td>
<ul>
<li>Are the design and/or development project stages defined? Where?</li>
<li>Are verification and validation activities addressed and appropriate?</li>
<li>Is it clear who is responsible for what?</li>
<li>Are the communication channels defined?</li>
<li>Does evidence exist to show that communication on projects is occurring and is effective?</li>
</ul>
</td><td>
Risk assessment should be utilized to assess the potential for, and the effect of, possible failures or faults in products or processes.
Management should consider the following factors that contribute to meeting the product and process performance expected by customers:
<ul>
<li>Life cycle</li>
<li>Safety and health</li>
<li>Testability and usability</li>
<li>User-friendliness</li>
<li>Dependability and durability</li>
<li>Ergonomics</li>
<li>Product disposal and the environment</li>
</ul>
Process Control Tip:
Designs should be created utilizing cross-functional teams (design, manufacturing and quality engineering, suppliers, customers, and so on) to ensure that product and process performance factors mentioned above are met and to ensure the best plan for manufacturability (reduces product cost/increases margin). (Refer to Sections 7.2.1, 7.2.2.)
Develop a standard new product development/ enhancement plan that identifies stakeholders and methods, tasks, timelines, dependencies, and resource constraints (if they exist).
Truly following QFD will assure proper translation of requirements and documentation of inputs. Following DFSS protocol assures compliance to this section. (Refer to Section 7.2.1.)
</td></tr>
</table>

7.3 Design and Development

7.3.1 Design and Development Planning

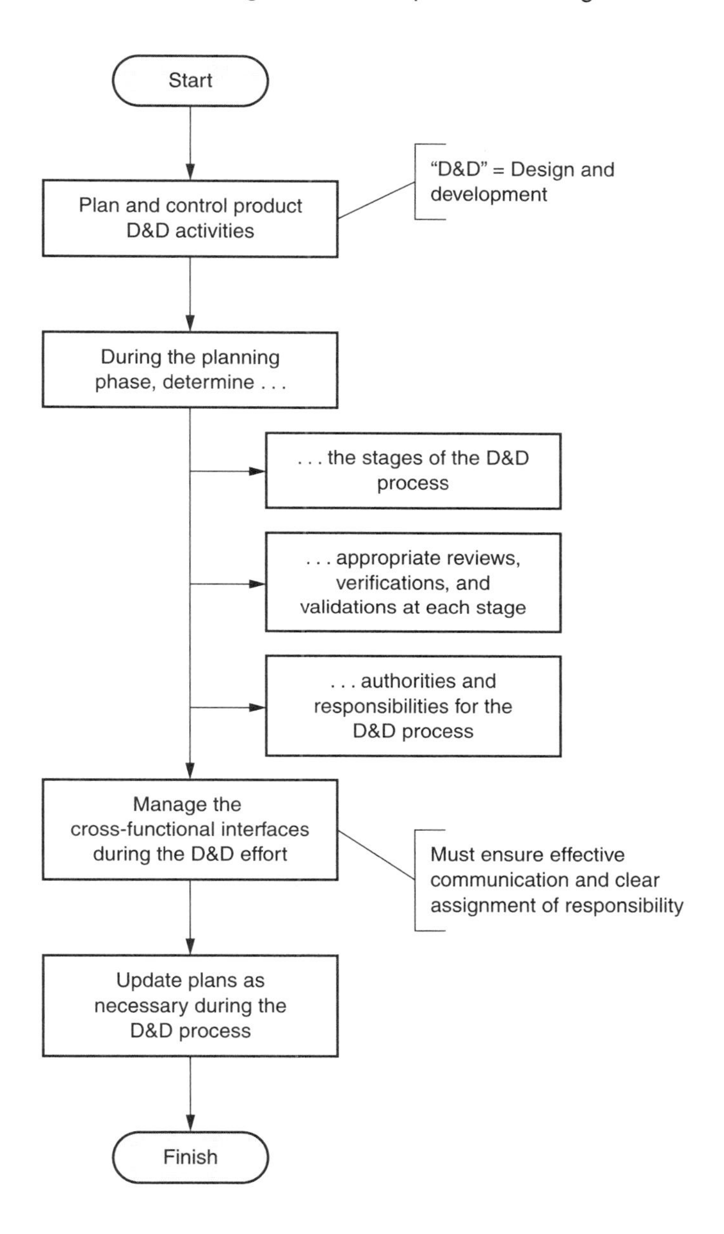

7.0 Product Realization

7.3 Design and Development

7.3.2 Design and Development Inputs

AS9100 The organization is to identify all inputs required for the design and development process, evaluate their adequacy for determining product design and development, and treat the identified inputs for a product as a quality record. Four specific types of inputs are required to be included, and the inputs identified must cover all requirements, be clearly defined, and can not create conflicting situations.

ISO 9004 Recommends identification of process as well as product inputs and coupling of internal needs and expectations with external ones. Presents examples of external inputs, internal inputs, and inputs that identify critical process and product characteristics.

Document Requirements:

Record

Internal Audit Questions:

- Are new product requirements defined and documented?
- Are the requirements complete, unambiguous, and without conflict?

Management Summary:

External and internal needs and expectations should be suitable for translation into input requirements for design and development processes.

Process Control Tip:

QFD analysis ensures that conflicting design criteria are identified. Key characteristics are derived from critical to quality/cost:

- ***Customer requirements***
- ***Organizational design requirements***
- ***Organizational manufacturing capabilities or methods***

7.3 Design and Development

7.3.2 Design and Development Inputs

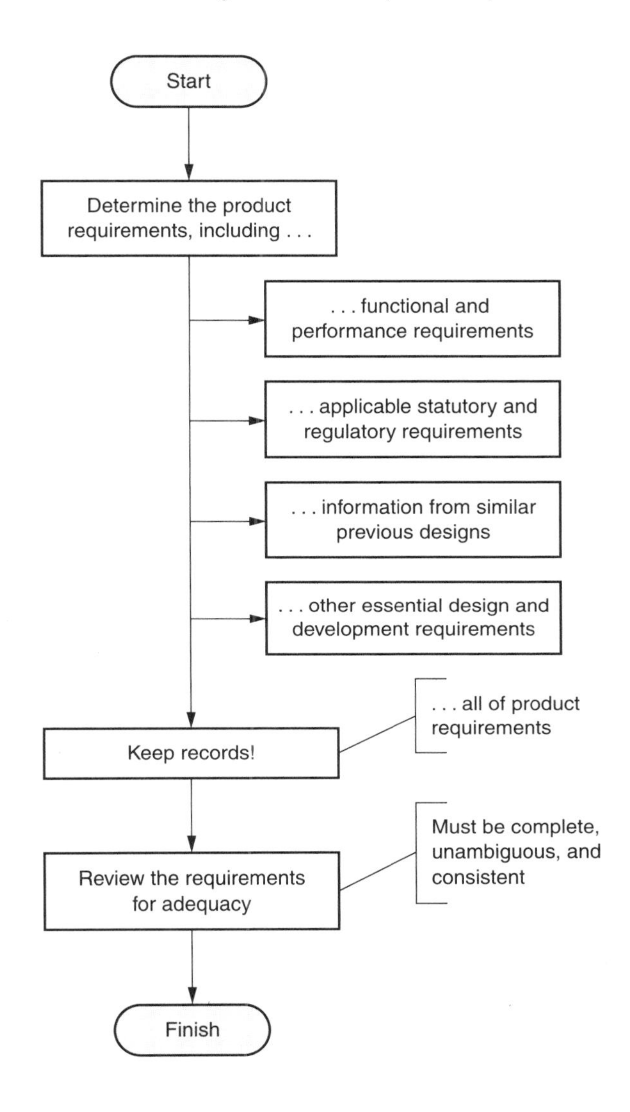

7.0 Product Realization

7.3 Design and Development

7.3.3 Design and Development Outputs

AS9100 The organization must capture and approve the outputs in a medium that permits verification against the inputs before their release. Design and development outputs must meet *five* criteria, including the ability to satisfy requirements identified as inputs, disseminate data to accomplish product-related processes within the organization, spell out product characteristics that are critical to its "safe and proper use," ***and identify key characteristics, when applicable, in accordance with design or contract requirements.***

Consideration must be given to proper inspection of features, identification and traceability, special processes, manufacturability, and configuration management.

ISO 9004 Provides examples of eight types of outputs.

Document Requirements:

None

Internal Audit Questions:

- Do approval processes exist for the release of products from the design and/or development process?
- Does a form with the output of design and/or development projects exist?
- Does it show how design and/or development outputs satisfy input requirements?
- Does output provide information for production operations?
- Are product acceptance criteria clearly stated?
- Are product safety and use characteristics identified?

Management Summary:

Design *and* development outputs should be reviewed against inputs to provide objective evidence that they have effectively met the process and product requirements.

Process Control Tip:

Key characteristics are noted on a design (usually owned by a prime) as a means to convey that those features are most critical to quality (CTQ). Typically a requirement for SPC data exists for key characteristics. Unfortunately, most shops send this data back to the customer and never use it themselves. If your process has a proven long-term C_{pk} that meets AS9103 (5.3), you can greatly reduce the inspection performed on these parts.

The sole intention of key characteristics is to monitor the CTQ characteristics through process control with feedback to design engineering for improved design and manufacturability. If the CTQs are controlled, typically the entire part is.

Ideally, the manufacturer and designer would perform analysis on the design to determine the CTC (critical to cost) characteristics. If the quality plan for product included this analysis, and the processes that generate the CTQs and CTCs were controlled processes, no other inspection would be required on the product.

7.3 Design and Development

7.3.3 Design and Development Outputs

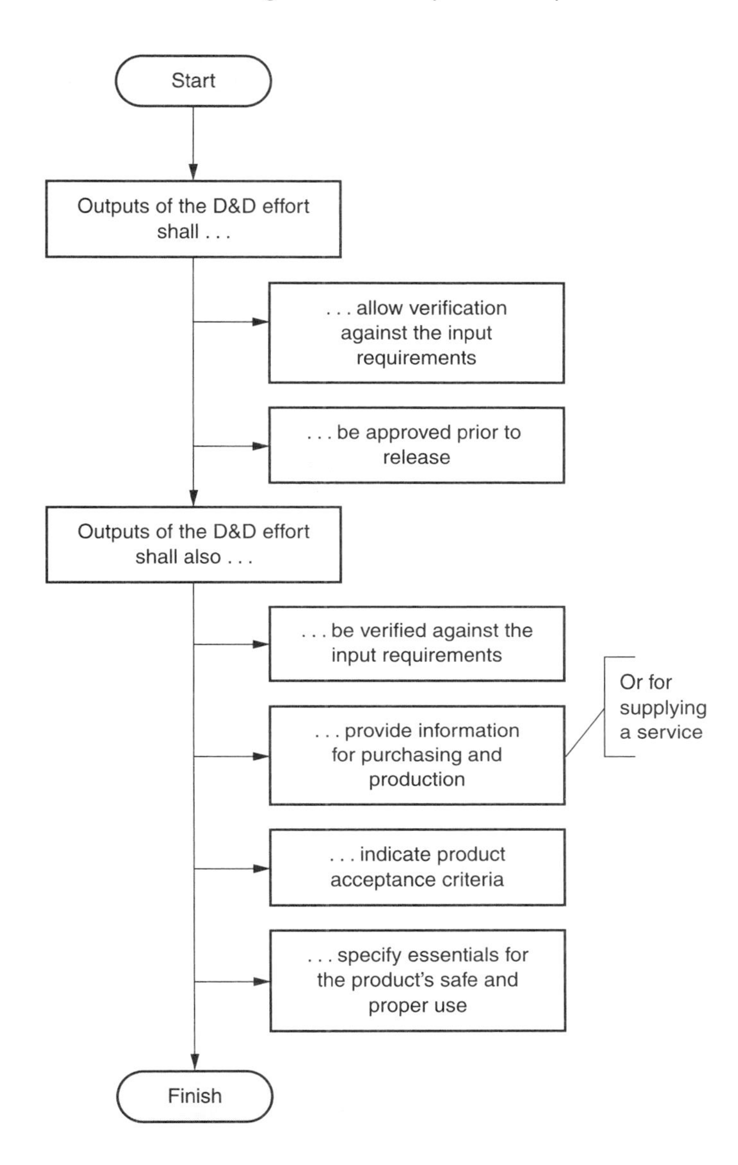

7.0 Product Realization

7.3 Design and Development

7.3.4 Design and Development Review

AS9100	The organization is to engage in "systematic reviews" of design and development activities at appropriate stages to ensure that the results are meeting product and design and development requirements and to catch and correct problems before they impact overall design and development. All groups involved in design and development should be represented in these reviews, and the review results, ***including authorization to progress to the next design stage,*** are to be treated as quality records.
ISO 9004	Provides examples of topics to be covered during the reviews.

Document Requirements:

Record

Internal Audit Questions:

- Are reviews of design and/or development being performed? Are they indicated in the project planning documents?
- Who attends these reviews? Is the attendance appropriate?
- Are results of the reviews documented? Are follow-up actions taken?

Management Summary:

Systematic reviews may be conducted at suitable stages in the design and development process as well as at completion.

Process Control Tip:

Well-documented designs that relate why decisions were made to date ensure that the designs are comprehensively reviewed.

QFD and/or a "requirements–traceability–acceptance" matrix are tools of choice to assure that design changes in one area do not adversely affect another designed feature or attribute.

7.3 Design and Development

7.3.4 Design and Development Review

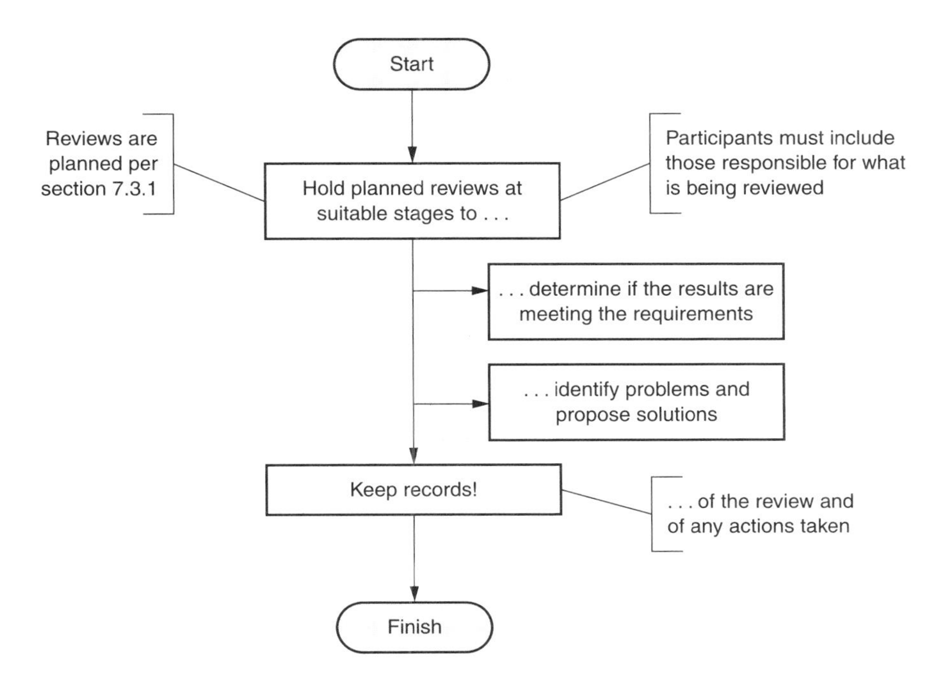

7.0 Product Realization

7.3 Design and Development

7.3.5 Design and Development Verification

AS9100 The organization must verify that design and development outputs meet all input requirements and that the results of verification activity are treated as quality records.

ISO 9004 Provides examples of verification and validation activities to be undertaken to ensure that the end product will be well received by customers and others.

ISO 9000 Defines "verification" in 3.8.4 as the "confirmation, through the provision of objective evidence, that specified requirements have been fulfilled," with activities that constitute confirmation provided in a Note.

Document Requirements:

Record

Internal Audit Questions:

- Is there a verification process in place?
- Is it effectively implemented?
- Are results of the verification documented?
- Are follow-up actions recorded?

Management Summary:

The organization should evaluate design and development outputs and processes in order to prevent nonconformities and deficiencies.

Process Control Tip:

The notes in AS9100 regarding this section pertain to alternative verification methods.

Brainstorm with a cross-functional team to find ways of verifying the design. Try to find ways to verify the design:

- ***Benchmark other related products***
- ***Utilize a source outside of the design team to work calculations to verify a result without the influence of the member involved in initial calculations.***

Use of tools aforementioned in 7.3.5 may assure that acceptance criteria meet the needs of your customer.

7.3 Design and Development

7.3.5 and 7.3.6 Design and Development Verification and Validation

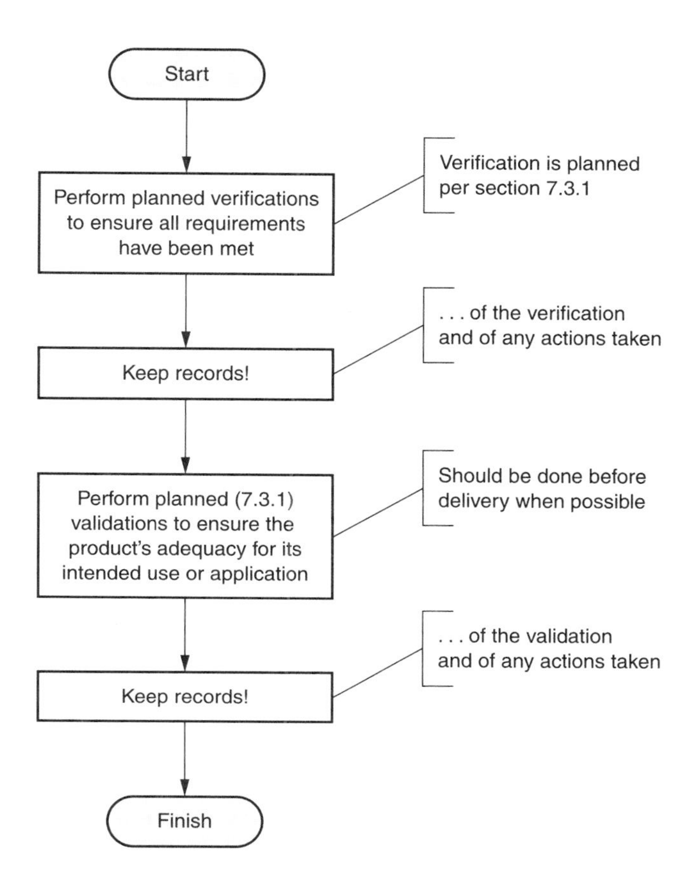

7.0 Product Realization

7.3 Design and Development

7.3.6 Design and Development Validation

AS9100 The organization must validate a product's ability to meet customer specifications and "known intended use or application" at the conclusion of the design and development process. Validation is to be conducted before product delivery or its "implementation" whenever it is possible to do so after the product's completion. Validation results and any necessary actions subsequent to the validation are to be treated as quality records.

- ***Notes in the standard imply that verification should occur prior to validation, and that validation(s) occur under the condition(s) intended for the use of the article. Take the opportunity to verify and validate throughout the design process.***

When the design phase is complete the organization shall be able to demonstrate that all work performed to arrive at the design is valid for meeting specifications.

Test plans are to be controlled, reviewed by the design team, and recorded to prove outcomes. Acceptance criteria must be defined and met.

ISO 9000 Defines "validation" in 3.8.5 as "confirmation, through . . . objective evidence, that the requirements of a specific intended use or application have been fulfilled."

Document Requirements:

Record

Internal Audit Questions:

- Is design and/or development validation performed to confirm that the product is capable of meeting requirements for intended use?
- Is validation completed prior to delivery when applicable?
- Is partial validation provided when full validation can not be performed prior to delivery?
- Are design validation results documented?
- Are follow-up actions recorded?
- ***Are design decisions and changes traceable?***
- ***Have acceptance criteria for the design been utilized?***

Management Summary:

Customers, suppliers, people in the organization should be able to use and evaluate the validation output prior to:

- Construction, installation, or application of engineering designs
- Installation or use of software outputs
- Widespread introduction of services

Process Control Tip:

Design trees like a "requirements to traceability to acceptance" (RTA) matrix will help show the opportunities for interim testing of "branches" of the design.

7.3 Design and Development

7.3.5 and 7.3.6 Design and Development Verification and Validation

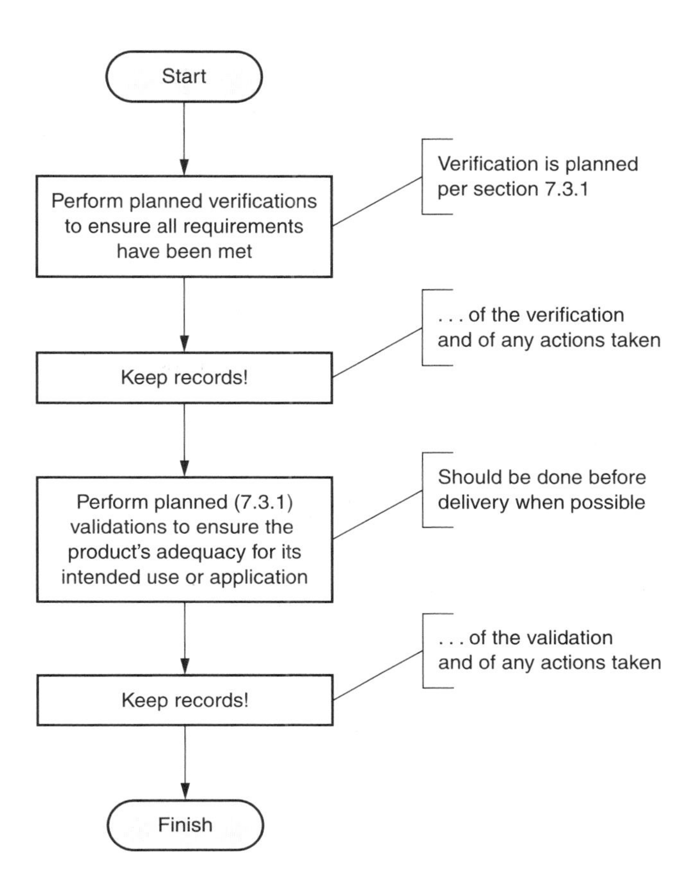

7.0 Product Realization

7.3 Design and Development

7.3.7 Control of Design and Development Changes

AS9100 When a change to a product's design and/or development occurs, it must be noted, with the information to be treated as a quality record. The organization must examine each change to ensure that the product will remain in conformance with customer and any other requirements before approving a change. The change's impact on both components and the product delivered to the customer must be covered in the examination. ***Ensure that regulatory or customer approval necessary for design changes occurs.***

Document Requirements:

Record

Internal Audit Questions:

- Are all design and/or development project changes documented?
- Is there evidence to demonstrate that changes are authorized?
- Do records include the results of review of changes?
- Have changes been communicated to interested parties?
- Do records include follow-up actions related to the review of changes?

Management Summary:

Verification and validation data should be reviewed by methods such as:

- Process and product improvement
- Output usability
- Adequacy of process and review records
- Investigation of failures
- Future design and development process needs

Process Control Tip:

Proper use of QFD, or RTA matrices, ensures that all customer and regulatory requirements stay connected to the applicable design specifications. This protects your organization from pursuing changes inadvertently without FAA or customer consent.

7.3 Design and Development

7.3.7 Control of Design and Development Changes

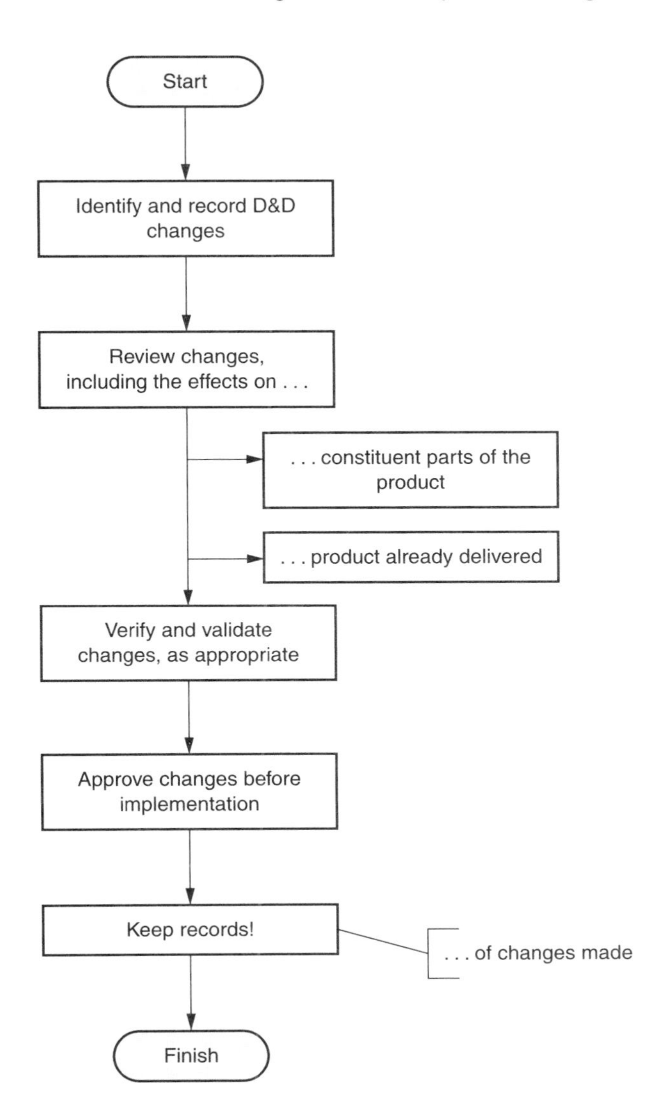

7.0 Product Realization

7.4 Purchasing

7.4.1 Purchasing Process

AS9100 The organization must make sure that materials, components, and services purchased from suppliers satisfy customer specifications for the organization's product, with the degree of supplier oversight proportional to the impact a part or service has on the product or processes. Suppliers must be qualified and capable of supplying parts/services needed to satisfy product and organizational requirements based on criteria developed by the organization for choosing suppliers and engaging in ongoing verification of qualifications and capability. The organization must treat as quality records ***an approved supplier list that addresses scope of approval,*** the results of its ***periodic*** verification activities ***that establish scope,*** and actions taken in response to verification outcomes. ***The organization is responsible for taking action when suppliers do not meet customer requirements or use customer-specified sources, including customer requirements, and this function should be controlled by a body with autonomy to disapprove sources when appropriate.***

ISO 9004 Recommends "electronic linkage" with suppliers to ensure communication of requirements and supplier participation in setting specifications for parts and services. Encourages consideration of 15 types of activity in the purchasing process. Provides 10 inputs organizations should use in managing supplier quality and advises organizations to assist suppliers in their development.

ISO 9000 Notes in 3.3.6 that suppliers include providers of a service or information.

Document Requirements:

Record

Internal Audit Questions:

- Have criteria for selection and periodic evaluation of suppliers been defined?
- Does a process exist for selecting and evaluating suppliers?
- Are evaluation results documented and retained as records?
- Are follow-up actions documented?

Management Summary:

Management should consider electronic communication of requirements to suppliers.

The organization should consider involving suppliers in the purchasing process to help the organization control inventory.

Process Control Tip:

The purchasing organization is often measured only by cost of material. Holding all areas of your organization accountable for quality and conformance better enables smart decisions when deciding to change sources.

Approved sources are best listed in a matrix where designs/parts and regulatory/customer requirements are traceable for use as critical criteria for purchase decisions.

7.4 Purchasing

7.4.1 Purchasing Process

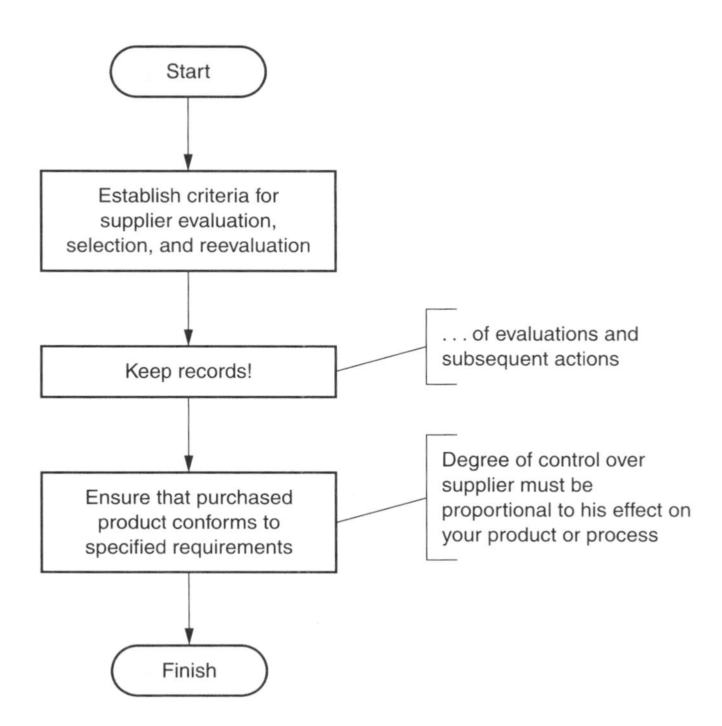

7.0 Product Realization

7.4 Purchasing

7.4.2 Purchasing Information

AS9100 The specifications for a product being purchased from a supplier must be part of the purchasing information, which must include approval requirements for the product and related items and activities, competency requirements for employees working on the product, QMS requirements for the supplier's operations, ***positive identification of data and specifications, and conveyance of verification, validation, and test specimen requirements. Notification of nonconforming material and, if applicable, approval by the customer to ship same. Requests for notification of changes to approved processes and approval of same, when applicable, should be present on purchase order documentation. The organization is required to make sure all requirements are identified and spelled out before transmitting the purchasing information to the supplier. Ensure that right-of-access by the organization or its customer or regulatory agencies is conveyed. All requirements listed above, including design key characteristics, that are applicable to sub-tier suppliers are to be communicated in purchase documentation.***

Document Requirements:

None

Internal Audit Questions:

- Does purchasing information adequately describe the products being ordered?
- Does purchasing information include (where appropriate) requirements for approval or qualification of product, procedures, processes, equipment, and personnel?
- Does purchasing information include (where applicable) quality management system requirements?
- Is purchasing information reviewed/approved to assure adequate descriptions of the specified requirements prior to release?
- ***Are purchase requirements communicated so that specifications, and so on, are well defined and unmistakable?***
- ***Are key characteristics identified to the supplier?***
- ***Does the purchasing documentation communicate applicable need for nonconforming material approvals and/or test specimens?***

Management Summary:

Management should consider the following activities when creating purchasing processes:

- Unique supplier processes
- Warranty replacement for nonconforming purchased products
- Logistic requirements
- Control of purchased product deviating from requirements
- Access to supplier's premises
- Mitigation of risks associated with purchased product

Process Control Tip:

Checklists (paper or automated within your purchasing system) help to error-proof documentation by ensuring that information is complete.

Requirements traceability matrices make it easier for the purchasing organization to assure completeness of requirements within purchasing documentation prior to release to suppliers.

7.4 Purchasing

7.4.2 Purchasing Information

7.4.3 Verification of Purchased Product

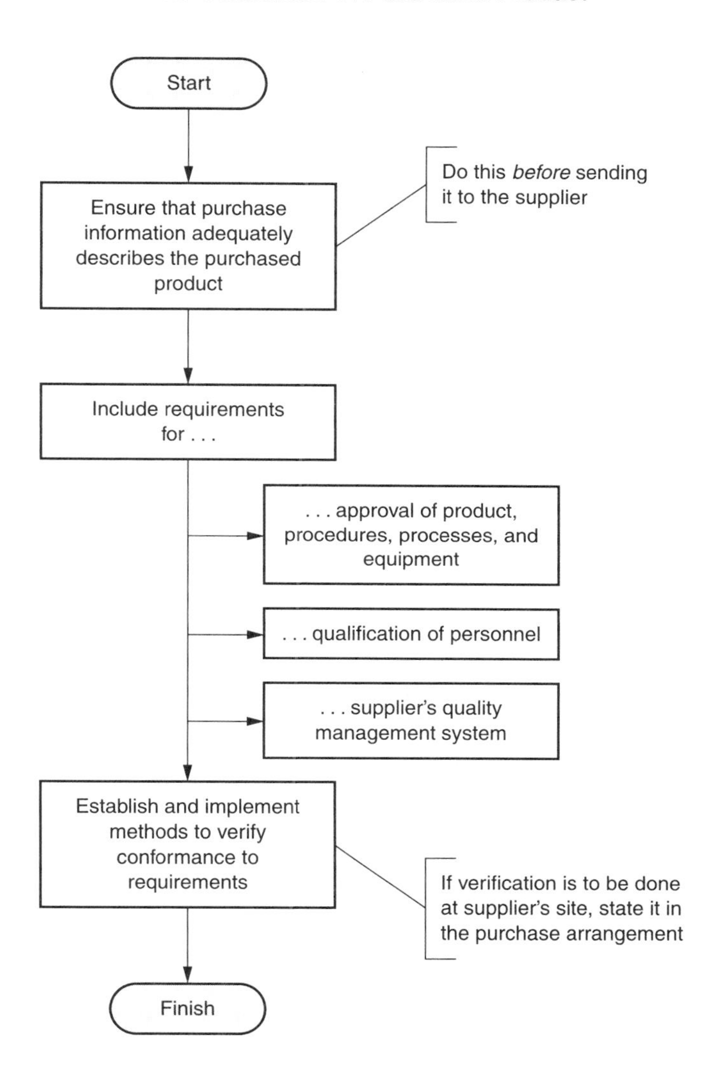

7.0 Product Realization

7.4 Purchasing

7.4.3 Verification of Purchased Product

AS9100 The organization is to create and use verification practices to confirm that purchased product conforms to the requirements conveyed to the supplier in the purchasing information, ***at any location necessary, although that verification does not release the supplier from liability regarding material condition and part conformance.*** The purchasing information also must indicate whether ***and how*** the organization or its customer will conduct on-site inspections of the product, processes, and/or supplier's facility and the related procedures for product release. ***The organization shall not utilize purchased product until its conformance is verified or it is controlled and released under positive recall. Periodic validation of raw material shall be performed by the organization or designated, with designations recorded.***

ISO 9000 Defines "release" in 3.6.9 as "permission to proceed to the next stage of a process."

Document Requirements:

None

Internal Audit Questions:

- Has the organization defined a process for verifying that purchased product conforms to defined requirements?
- Is the process effectively implemented?
- Does objective evidence exist of product acceptance?
- Is verification of purchased product performed at the supplier's premises?
- If so, are the arrangements specified, and does objective evidence exist to show effective implementation?
- ***Is there evidence of a system that controls material determined to be nonconforming and allows it to be traced through the system?***

Management Summary:

In the event of supplier failure, management should have actions in place to maintain the organization's performance.

Process Control Tip:

The opportunity for standard quality planning (internal or purchased material) becomes possible when your organization becomes AS9100 compliant.

Prior to AS9100, organizations and their suppliers had to plan individual purchase orders through the contracted process.

Today, if all contracts are processes within an AS9100 system, the huge majority—if not all—of the PO quality requirements are covered by the system. Some primes have already identified in their requirement flow-down only those requirements above and beyond AS1900.

Create a matrix of AS9100 to prime flow-down requirements or ask the customer for this information to save time and resources.

7.4 Purchasing

7.4.2 Purchasing Information

7.4.3 Verification of Purchased Product

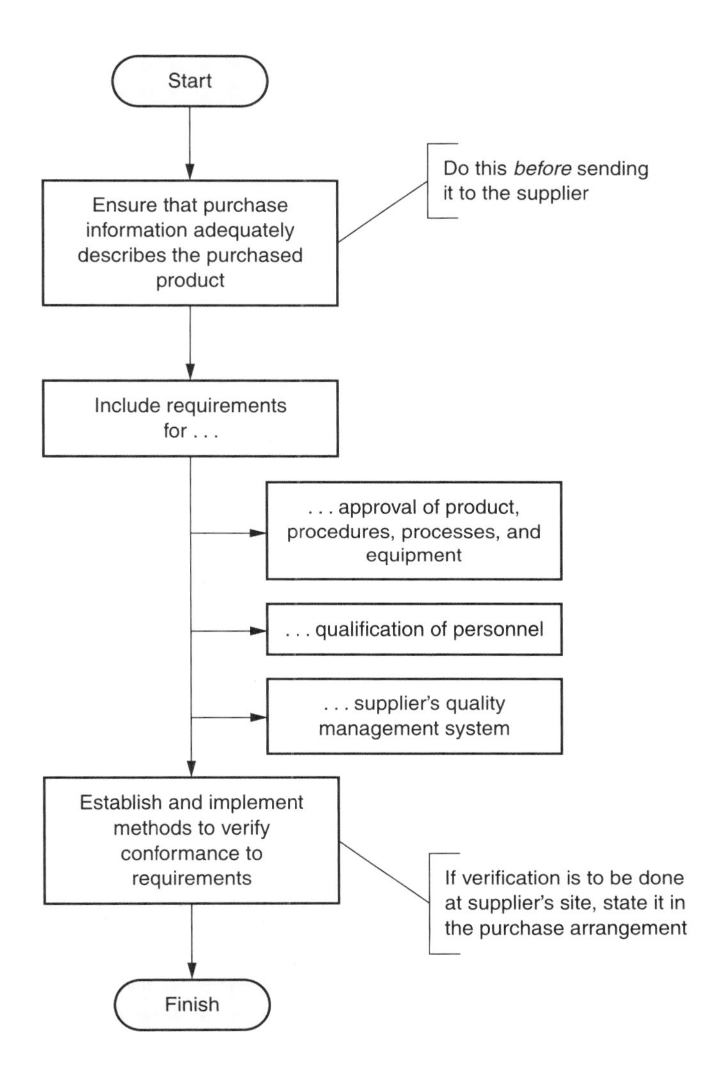

7.0 Product Realization

7.5 Production and Service Provision

7.5.1 Control of Production and Service Provision

AS9100 Production and service operations are to be conducted under planned procedures that control the activities involved, including production, monitoring and measuring, delivery, and post-delivery servicing. To ensure that processes and products conform to customer specifications and other requirements, the organization is required to control six items, including the availability and/or use of procedures, work instructions, information detailing product characteristics, and proper equipment. ***Establish standard operating procedures, tracking mechanisms, process controls, and control plans for key characteristics, NC programs, and outside processing—including special processes—that include verification of conformance at the first point possible in the manufacturing process. Change control is required. Plan for opportunities to record all necessary measurements, using variable data (not attribute—go/no-go) for key characteristics. The process shall ensure that all product be accounted for, and is protected from foreign object damage (FOD). Infrastructure should be controlled to assure good quality, including storage and control of tooling.***

ISO 9004 Provides guidance on going beyond control of production and service processes to increased effectiveness and efficiency and associated activities that support these processes, with examples of the support activities provided.

ISO 9000 Defines "quality control" in 3.2.10 as "part of quality management, focused on fulfilling quality requirements."

7.5.1.4 Post-Delivery Support

AS9100 To ensure that processes meet the standard, the organization shall make provisions for collection and analyzing of in-service data and actions taken based on problems identified, technical document control, and control of repair schemes and off-site work.

Document Requirements:

None

Internal Audit Questions:	*Management Summary:*
• Are specifications available that define quality characteristic requirements of the product and/or service? • Has the organization demonstrated the suitability of equipment for production and service operations to meet product and/or service specifications? • Are all production and service operations that require control defined (including those that need ongoing monitoring, work instructions, and/or special controls)? • Are work instructions available and adequate to permit control of the appropriate operations to ensure conformity of product and/or service? • Have the work environment requirements been defined and are they being met to ensure conformity of the product and/or service? • Is suitable measuring and monitoring equipment available when and where necessary to ensure conformity of the product and/or service? • Have monitoring and verification activities been planned and are they carried out as required? • Have suitable processes for hardware, processed material, and software been implemented for release of the product and for their delivery to the customer? • Are suitable release mechanisms in place to ensure that service conforms to requirements?	To improve the effectiveness and efficiency of the realization processes, management should consider the following: • Reducing waste • Developing supplier capability • Improving infrastructure • Processing methods and process yield • Monitoring methods ***Process Control Tip:*** ***Control plans are the core of process control. The standard is implying that implementing a control plan assures a better process. More work up front needs to be performed to formulate control plans, but they assure quality and cost control for the long term.*** ***The basic process for creating a control plan includes product and process planning, identification of key process and product measures (CTCs, CTQs), risk analysis (that is, FMEA), and development of control plans—including reaction (change control) plans. Control plans can include provisions for the use of covers, and so on, to mitigate the risk of FOD.*** ***A lean manufacturing environment ensures that all tooling is accounted for and that cleanliness is upheld—also reducing the risk of quality problems, contamination, and FOD. Lean also promotes the use of visual systems for process control and visual product verification purposes to error-proof the processing of product.***

7.5 Production and Service Provision

7.5.1 Control of Production and Service Provision

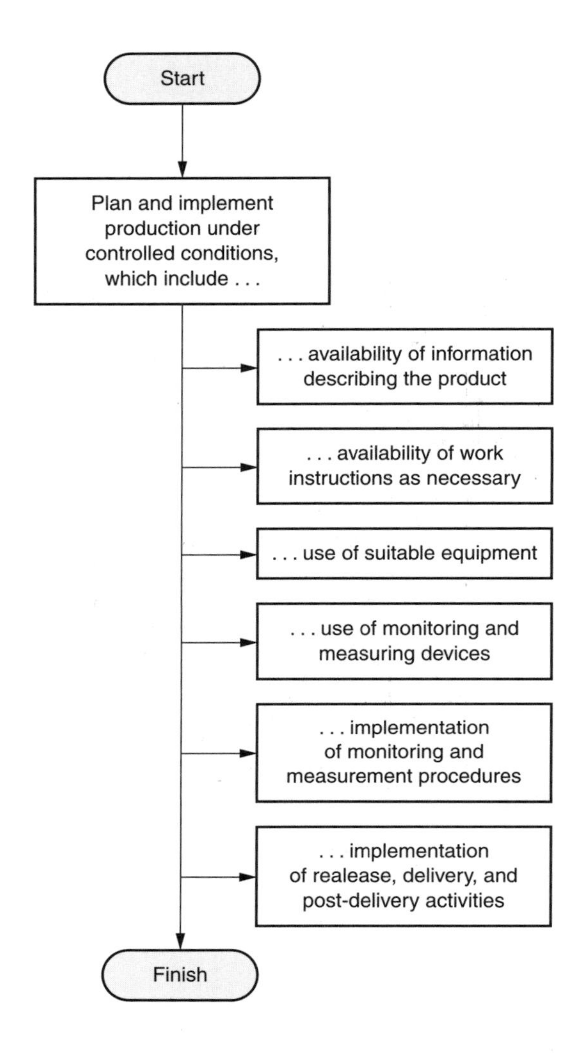

7.0 Product Realization

7.5 Production and Service Provision

7.5.2 Validation of Processes for Production and Service Provision

AS9100 When a product or service can not be tested during and/or after production and/or service delivery to verify its conformance to customer and other requirements, the organization must ensure that the processes involved are capable of producing a product or service that conforms to requirements, validating the ***(special)*** processes. There are six items the organization must have in place to ensure that processes are validated and remain in conformance, including criteria for process ***qualification,*** review and approval, employee and equipment approval procedures, and ***specified*** production/service techniques and processes to be followed.

Document Requirements:

Record (d)

Internal Audit Questions:

- Are there defined criteria for review and approval of the validation of processes?
- Is there a process in place to handle deficiencies with product that is in use or that has been delivered?
- Are there records of validation of processes for production and service provision?

Management Summary:

The organization must also validate processes when product that is in use or has been delivered results in deficiencies.

Process Control Tip:

Special processes are controlled processes—the key parameters are controlled and monitored. Since material that undergoes a special process can not be validated unless destructively tested, the process is strongly adhered to.

Special processing is the ultimate process control. The process CTQs are understood, documented, and controlled to assure 100% quality. Any process can be controlled in such a manner to ensure that product coming out of the process (turning, machining, and so on) is conforming. This is the secret to reducing redundant inspection—understanding and controlling the processes that affect fit, form, or function of product.

7.5 Production and Service Provision

7.5.2 Validation of Processes for Production and Service Provision

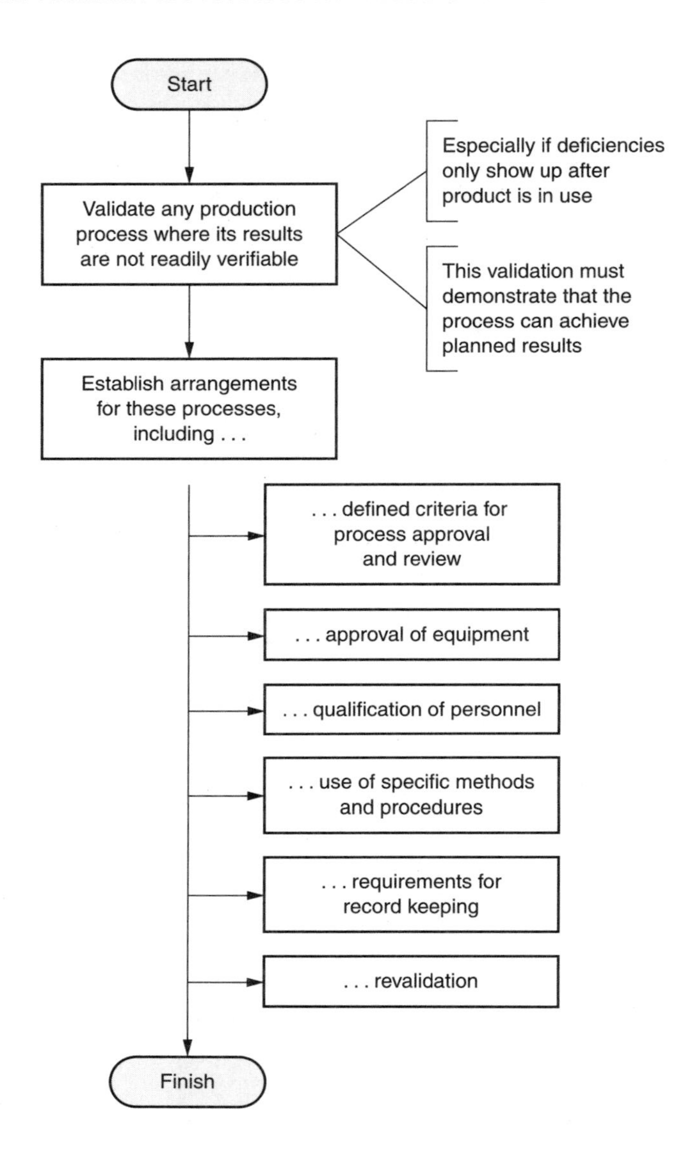

7.0 Product Realization

7.5 Production and Service Provision

7.5.3 Identification and Traceability

AS9100 The organization must have an effective system in place to maintain identification of a product throughout the production process, including its inspection and test status, whenever these are needed. When a product must be traceable, the organization must impose a "unique identification" on each product and record the identification information to permit traceability, with the record to be treated as a quality record. ***Traceability of any variance from original design, including rework or authorized repair, must be enforced. Controls of authorization and identification media must exist. According to customer or regulatory requirements, identification of product (and therefore the records of traceability) may be required for the life of the product, with lot traceability and record of production from raw material to delivery of product or destruction of scrap.***

ISO 9004 Recommends collecting data during the identification and traceability processes to go beyond product requirements to improving the product and its processes.

ISO 9000 Notes in 3.5.4 that traceability can relate to the origin of materials and parts, among other things.

Document Requirements:

Record

Internal Audit Questions:

- Has the product been identified by suitable means throughout production and service operations?
- Has the status of the product been identified at suitable stages with respect to measurement and monitoring requirements?
- Is traceability a requirement? If so, is the unique identification of the product recorded and controlled?

Management Summary:

In order to collect data for improvement, the organization may establish a process for identification and traceability that goes beyond regular requirements.

Process Control Tip:

Use visual controls to tag product that separates from its parent lot and to trace accompanying paperwork.

7.5 Production and Service Provision

7.5.3 Identification and Traceability

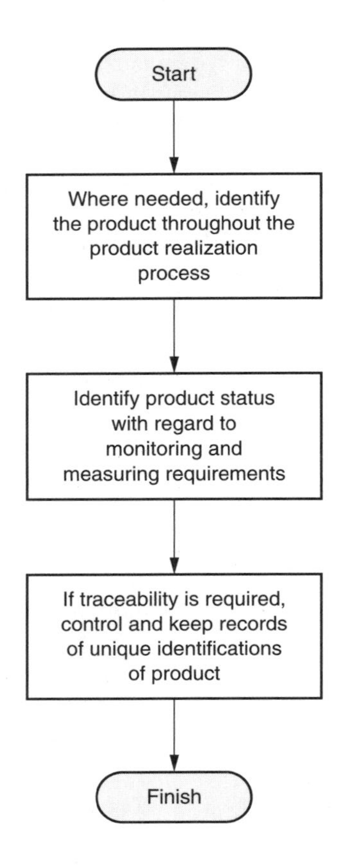

7.0 Product Realization

7.5 Production and Service Provision

7.5.4 Customer Property

AS9100 The organization is to have procedures to protect from damage, misuse, or loss property supplied by the customer for use by the organization in its processes or as components in the product to be supplied to the customer, including intellectual property ***(data for design, production, or inspection).*** The organization must alert the customer if its property is nonconforming and keep records of these incidences as quality records.

ISO 9004 Provides examples of the types of property customers or other interested parties might entrust to an organization for use or in production/service activities.

Document Requirements:

Record

Internal Audit Questions:

- Has the organization identified, verified, protected, and maintained customer property provided for incorporation into the final product?
- Does control of product extend to all customer property, including intellectual property?
- Are there records that indicate when customer property has been lost, damaged, or otherwise found to be unsuitable?
- When customer property has been lost, damaged, or otherwise found to be unsuitable, is there evidence that the customer has been informed?

Management Summary:

Examples of customer property that the organization may be responsible for include:

- Ingredients or components for inclusion in a product
- Product supplied for repair, maintenance, or upgrading
- Packaging materials
- Customer property transported to a third party
- Customer intellectual property
- Tooling
- Gauging
- Designs
- Software code

Process Control Tip:

Traceability becomes important for customer-supplied material, tooling, or intellectual property. At any time, the customer may choose to validate that measures are put in place to control and maintain supplied articles and data. Simple visual systems and electronic or physical marking on data files, documents, software code, tooling, and so on, can provide easy traceability and assure that errors will not be made in the maintenance of customer property.

7.5 Production and Service Provision

7.5.4 Customer Property

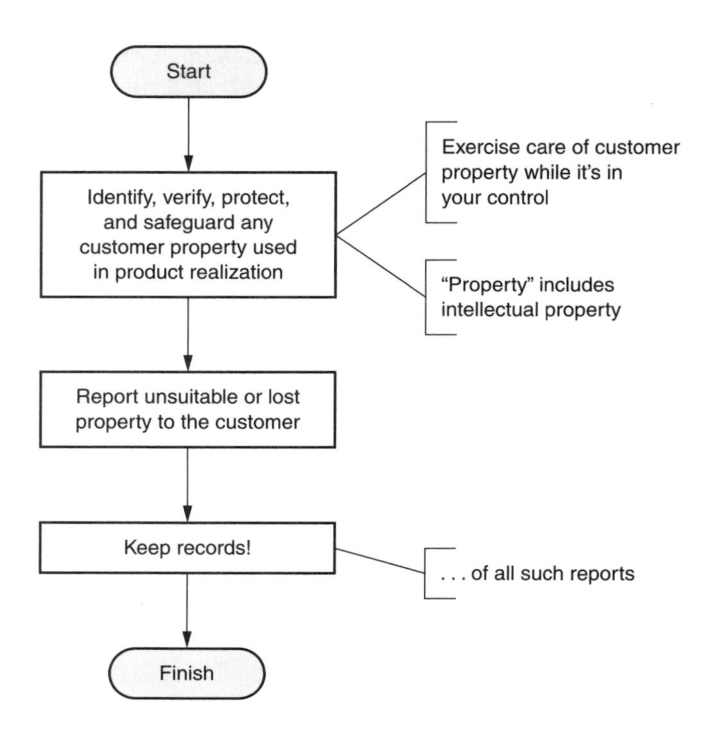

7.0 Product Realization

7.5 Production and Service Provision

7.5.5 Preservation of Product

AS9100 The organization must ensure that a product and its components continue to satisfy customer specifications and other requirements throughout the production process until delivery to the customer's intended operations.

Preservation of product and documents shall be adhered to.

ISO 9004 Advises management to ensure that resources are available to protect a product from destruction, deterioration, or misuse while in the organization's possession and to inform the customer and others of what is required to preserve the product during its life cycle.

Document Requirements:

None

Internal Audit Questions:

- Is product identified during internal processing and delivery?
- When handling product during internal processing and delivery, does the organization preserve conformity to customer requirements?
- When packaging product during internal processing and delivery, does the organization preserve conformity to customer requirements?
- When storing product during internal processing and delivery, does the organization preserve conformity to customer requirements?
- Does the organization protect the product during internal processing and delivery to preserve conformity to customer requirements?

Management Summary:

To control handling, packaging, storage, preservation, and delivery of product, management should consider special requirements that may arise from the nature of such products as:

- Software
- Electronic media
- Hazardous materials
- Products requiring special service people
- Unique or irreplaceable products and materials

7.5 Production and Service Provision

7.5.5 Preservation of Product

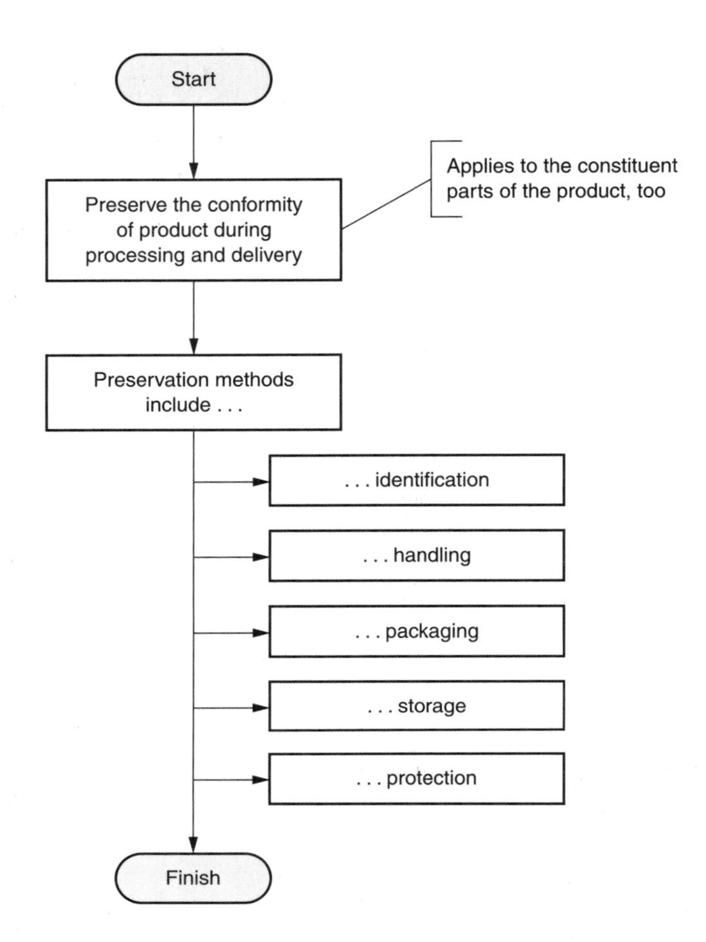

7.0 Product Realization

7.6 Control of Monitoring and Measuring Devices

AS9100 The organization is to decide what types and frequency of monitoring and measuring are required to ensure that a product satisfies customer and other requirements, identify the devices necessary for these activities, and put in place procedures to ensure that monitoring and measurement is conducted in accordance with the organization's specifications. The organization must maintain ***and register*** the devices in a condition to ensure accurate monitoring and measurements, including calibration, with the results, along with device maintenance records, to be treated as quality records. Specific requirements are spelled out for computer software used for monitoring and measurement purposes. When a device is out of conformity, previous measurements must be examined to ensure that the nonconformity did not result in product nonconformities. ***This includes any equipment that gauges or reports test or inspection data.***

ISO 9004 Expands on what monitoring and measurement processes could involve and achieve, with emphasis on the organization taking steps to eliminate potential problems within processes and thus reduce the need to rely on monitoring and measuring activities.

ISO 9000 Provides related definitions in 3.8.2, Inspection, and subsection 3.10, Terms Related to Quality Assurance for Measurement Processes.

Document Requirements:

- Record (a)
- Record
- Record

Internal Audit Questions:	Management Summary:
• Has the organization identified the measurements to be made? Has the organization identified the measurement and monitoring devices required to assure conformity of product to specified requirements? • Are measuring and monitoring devices used to ensure measurement capability? Are they calibrated and adjusted periodically or prior to use against devices traceable to international or national standards? Are those calibration results recorded? • When traceability to international or national standards can not be done since no standards exist, is the calibration recorded? • Are measuring and monitoring devices safeguarded from adjustments that would invalidate the calibration? Are they protected from damage and deterioration during handling, maintenance, and storage? • Does the organization have the validity of previous results from measuring and monitoring devices reassessed if they are subsequently found to be out of calibration? Is corrective action taken? • Is software used for measuring and monitoring of specified requirements validated prior to use?	For data confidence, measuring and monitoring processes should include confirmation that the device is fit for use, identifiable, and maintained to suitable accuracy and accepted standards.

7.6 Control of Monitoring and Measuring Devices

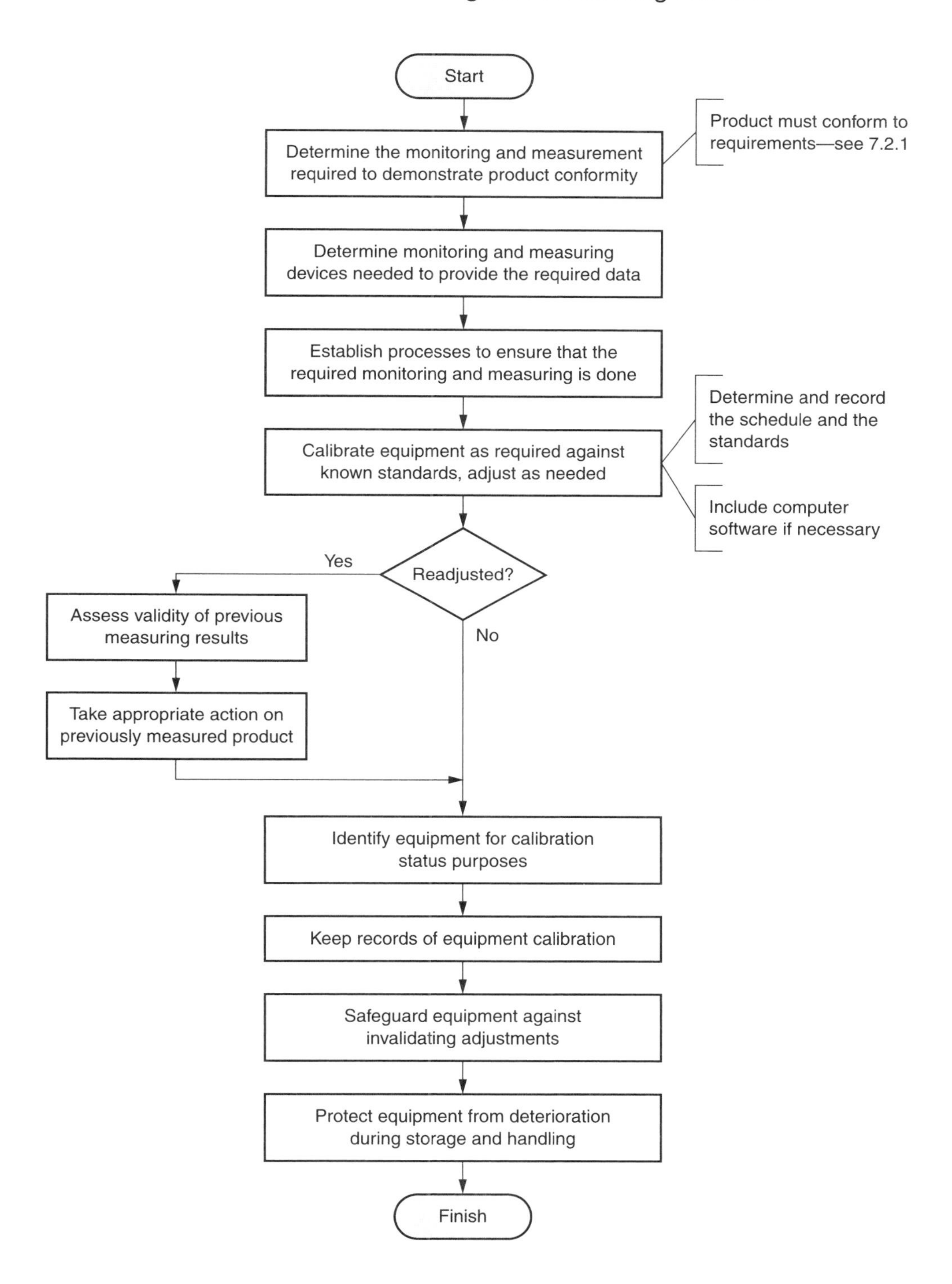

REVISIONS TO AS9100:2009 (REVISION C), CLAUSE 7

(Source: Larry Whittington)

7. Product Realization

7.1 Planning of Product Realization

AS9100C added this Note under 7.1, subclause (a) on what to consider for product quality objectives:

> *In planning product realization, the organization shall determine the following, as appropriate:*
>
> *a) quality objectives and planning for the product;*
>
> *Note: Quality objectives and requirements for the product include consideration of aspects such as*
>
> *- product and personal safety,*
> *- reliability, availability, and maintainability,*
> *- producibility and inspectability,*
> *- suitability of parts and materials used in the product,*
> *- selection and development of embedded software, and*
> *- recycling or final disposal of the product at the end of its life.*

The ISO 9001:2008 portion of AS9100C added "measurement" as one of the required activities to be determined during the planning of product realization:

> *In planning product realization, the organization shall determine the following, as appropriate:*
>
> *b) the need to establish processes~~,~~ and documents, and to provide resources specific to the product;*
>
> *c) required verification, validation, monitoring, measurement, inspection, and test activities specific to the product and the criteria for product acceptance;*

AS9100C added a new requirement at subclause (e) and moved the old (e) to become the basis for the new subclause (f). Remember, configuration management moved from clause 4.3 to clause 7.1.3.

> In planning product realization, the organization shall determine the following, as appropriate:
>
> *(e) configuration management appropriate to the product;*
>
> *(f) ~~the identification of~~ resources to support the use ~~operation~~ and maintenance of the product.*

7.1.1 Project Management

AS9100C added this new clause on project management and acceptable risk (see new clause 7.1.3), which adds focus on up-front planning and the ongoing management of project plans.

> *As appropriate to the organization and the product, the organization shall plan and manage product realization in a structured and controlled manner to meet the requirements at acceptable risk, within resource and schedule constraints.*

7.1.2 Risk Management

AS9100C added this new clause on risk management. In AS9100B "risk" was only mentioned at clause 7.2.2 (d) on reviewing product requirements for risks such as new technology and short delivery times. Placement in this clause will cause more focus on product risk during product realization:

> *The organization shall establish, implement, and maintain a process for managing risk to the achievement of applicable requirements, that includes as appropriate to the organization and the product:*
>
> *a) assignment of responsibilities for risk management,*
>
> *b) definition of risk criteria (e.g., likelihood, consequences, risk acceptance),*
>
> *c) identification, assessment, and communication of risks throughout product realization,*
>
> *d) identification, implementation, and management of actions to mitigate risks that exceed the defined risk acceptance criteria, and*
>
> *e) acceptance of risks remaining after implementation of mitigating actions.*

7.1.3 Configuration Management

AS9100C moved this clause from 4.3 to 7.1.3 and expanded it to describe what might be included in a configuration management process. Placement at this clause is to focus configuration management on the product and to sustain that focus throughout product realization.

> *The organization shall establish, ~~document~~ implement, and maintain a configuration management process that includes, as appropriate to the product:*
>
> *a) configuration management planning,*
>
> *b) configuration identification,*
>
> *c) change control,*
>
> *d) configuration status accounting, and*
>
> *e) configuration audit.*
>
> *NOTE: ~~Guidance on configuration management is given in~~ See ISO 10007 for guidance.*

7.1.4 Control of Work Transfers

AS9100C moved this requirement for the control of work transfers from clause 7.5.1.4 to 7.1.4. The rewording clarifies the types of work transfers and points out that it could be temporary or permanent.

> *~~Control of Work Transferred, on a Temporary Basis, Outside the Organization's Facilities:~~*
>
> *~~When planning to temporarily transfer work to a location outside the organization's facilities, the organization shall define the process to control and validate the quality of the work.~~*
>
> *The organization shall establish, implement, and maintain a process to plan and control the temporary or permanent transfer of work (e.g., from one organization facility to another, from the organization to a supplier, from one supplier to another supplier) and to verify the conformity of the work to requirements.*

7.2 Customer-Related Processes

7.2.1 Determination of Requirements Related to the Product

The change below from "related" to "applicable" shifts from determining legal requirements that are merely associated with the product to those that are relevant and can be applied to the product:

> *The organization shall determine:*
>
> *c) statutory and regulatory requirements ~~related~~ applicable to the product, and*

Since the bulleted list for 7.2.1 begins with "The organization shall determine," the use of the word "determined" again in the entry below was not appropriate. The new text clarifies that the additional requirements "considered necessary" must be determined.

> *d) any additional requirements ~~determined~~ considered necessary by the organization.*

AS9100C added this new Note. See clause 3.2 of AS9100C for the definition of "special requirements."

> *NOTE: Requirements related to the product can include special requirements.*

Organizations may not have considered the breadth of post-delivery activities as described by the new Note below.

> *NOTE: Post-delivery activities include, for example, actions under warranty provisions, contractual obligations such as maintenance services, and supplementary services such as recycling or final disposal.*

7.2.2 Review of Requirements Related to the Product

AS9100C added a new requirement at subclause (d) on special requirements, and moved the old subclause (d) requirement on risks to a new subclause (e).

> *This review . . . and shall ensure that*
>
> *d) special requirements of the product are determined, and*
>
> *e) ~~d)~~ risks (e.g., new technology, short delivery time ~~scale~~ frame) have been ~~evaluated~~ identified (see 7.1.2).*

7.2.3 Customer Communication

No changes in AS9100C clause 7.2.3.

7.3 Design and Development

7.3.1 Design and Development Planning

AS9100C removed the entry under subclause (a) since the topics are addressed in other sections:

> During the design and development planning, the organization shall determine
>
> a) the design and development stages,
>
> ~~*– in respect of organization, task sequence, mandatory steps, significant stages, and method of configuration control,*~~

AS9100C deleted text on design planning considerations and restated it in a new sentence:

> *Where appropriate, the organization shall divide the design and development effort into distinct activities and, for each activity, define the tasks, necessary resources, responsibilities, design content, input and output data, and planning constraints.*
>
> ~~*Where appropriate, due to complexity, the organization shall give consideration to the following activities:*~~
>
> ~~*- structuring the design effort into significant elements;*~~
>
> ~~*- for each element, analyzing the tasks and the necessary resources for its design and development. This analysis shall consider an identified responsible person, design content, input data, planning constraints, and performance conditions. The input data specific to each element shall be reviewed to ensure consistency with requirements.*~~

AS9100C moved the following text from the end of clause 7.3.1 to the middle of clause 7.3.1:

> *The different design and development tasks to be carried out shall be ~~defined according to specified~~ based on the safety ~~or~~ and functional objectives of the product in accordance with customer, statutory and~~/or~~ regulatory ~~authority~~ requirements.*

AS9100C added this requirement to have product designers consider more than just product function:

> *Design and development planning shall consider the ability to produce, inspect, test, and maintain the product.*

Clause 7.3.1.b continues to state that the organization must determine the review, verification, and validation appropriate for each design and development stage. The new Note below explains that although review, verification, and validation have distinct goals, they can be carried out separately or in any combination.

> *NOTE: Design and development review, verification, and validation have distinct purposes. They can be conducted and recorded separately or in any combination, as suitable for the product and the organization.*

7.3.2 Design and Development Inputs

This clause continues to require the design and development inputs to be determined and records to be maintained. It lists several types of requirements to be included. The revision below simply changes from "These inputs" to "The inputs."

> ~~*These*~~ *The inputs shall be reviewed for adequacy. Requirements shall be complete, unambiguous, and not in conflict with each other.*

7.3.3 Design and Development Outputs

The change below removes the unnecessary word "provided." It also switches from "a form that enables verification" to "a form suitable for verification." To enable something is to make it possible. However, to be suitable means it is meant for use, or in this case, for verification.

> *The outputs of design and development shall be ~~provided~~ in a form ~~that enables~~ suitable for verification against the design and development input and shall be approved prior to release.*

The change below was to simply remove the word "for."

> *b) provide appropriate information for purchasing, production, and ~~for~~ service provision,*

AS9100C expanded subclause (e) to include "critical items" (see definition in clause 3.3), not just "key characteristics" (see definition in clause 3.4):

> *e) ~~identify~~ specify, as applicable, any critical items, including any key characteristics, ~~when applicable, in accordance with design or contract requirement~~ and specific actions to be taken for these items.*

AS9100C revised this text at the end of section 7.3.4:

> *The organization shall define the ~~All pertinent~~ data required to allow the product to be identified, manufactured, inspected, used, and maintained ~~shall be defined by the organization~~; including for example:*
>
> *~~- drawings, part lists, specifications;~~*
>
> *- ~~a listing of those~~ the drawings, part lists, and specifications necessary to define the configuration and the design features of the product; and*
>
> *- ~~information on~~ the material, process~~es~~, ~~type of manufacturing~~ and assembly data needed ~~of the product necessary~~ to ensure ~~the~~ conformity of the product.*

The new Note below reminds the reader that clause 7, Production and Service Provision, includes subclause 7.5.5, Preservation of Product. Why do that? To indicate that the design output should consider product preservation, for example, product packaging.

> *NOTE: Information for production and service provision can include details for the preservation of product.*

7.3.4 Design and Development Review

No changes in AS9100C clause 7.3.4.

7.3.5 Design and Development Verification

AS9100C deleted the Note from this section. Since Notes are added for clarification, they can be deleted when they are no longer necessary.

> *~~NOTE: Design and/or development verification may include activities such as:~~*
>
> *~~- performing alternative calculations,~~*
>
> *~~- comparing the new design with a similar proven design, if available,~~*
>
> *~~- undertaking tests and demonstrations, and~~*
>
> *~~- reviewing the design stage documents before release.~~*

7.3.6 Design and Development Validation

AS9100C deleted the Notes from this section. Since Notes are added for clarification, they can be deleted when they are no longer necessary.

~~*NOTES:*~~

~~*- Design and/or development validation follows successful design and/or development verification.*~~

~~*- Validation is normally performed under defined operating conditions.*~~

~~*- Validation is normally performed on the final product, but may be necessary in earlier stages prior to product completion.*~~

~~*- Multiple validations may be performed if there are different intended uses.*~~

7.3.6.1 Design and Development Verification and Validation Testing

AS9100C did not change the content moved to this section from the old clause 7.3.6.2.

7.3.6.2 Design and Development Verification and Validation Documentation

AS9100C did not change the content moved to this section from the old clause 7.3.6.1.

7.3.7 Control of Design and Development Changes

ISO 9001:2008 merged the first and last paragraph of this section into one paragraph with no text changes. AS9100C added the sentence below:

Design and development changes shall be controlled in accordance with the configuration management process (see 7.1.3).

AS9100C deleted the following text since the new requirement in clause 4.1 requires that applicable statutory and regulatory requirements be addressed throughout the quality management system.

~~*The organization's change control process shall provide for customer and/or regulatory authority approval of changes, when required by contract or regulatory requirement.*~~

7.4 Purchasing

7.4.1 Purchasing Process

AS9100C modified the second paragraph in this section for clarity reasons:

The organization shall be responsible for the ~~quality~~ conformity of all products purchased from suppliers, including ~~customer-designated~~ product from sources defined by the customer.

AS9100C revised the section below to give examples of approval status and approval scope:

a) maintain a register of ~~approved~~ its suppliers that includes approval status (e.g., approved, conditional, disapproved) and the scope of the approval (e.g., product type, process family);

AS9100C revised the section below to replace the use of "records" with "results":

b) periodically review supplier performance; ~~records~~ the results of these reviews shall be used as a basis for establishing the level of controls to be implemented;

AS9100C rewrote and expanded the requirement in subclause (e), and added a requirement in (f) in keeping with the new focus on risk management.

> *~~e) ensure that the function having responsibility for approving supplier quality systems has the authority to disapprove the use of sources.~~*
>
> *e) define the process, responsibilities, and authority for the approval status decision, changes of the approval status, and conditions for a controlled use of suppliers depending on the supplier's approval status, and*
>
> *f) determine and manage the risk when selecting and using suppliers (see 7.1.2)*

7.4.2 Purchasing Information

AS9100C revised 7.4.2 subclauses (d) and (e). There were no changes to subclauses (a), (b), (c), or (f).

> Purchasing information shall describe the product to be purchased, including where appropriate
>
> *d) the ~~name or other positive~~ identification and revision status~~, and applicable issues~~ of specifications, drawings, process requirements, inspection/verification instructions, and other relevant technical data,*
>
> *e) requirements for design, test, ~~examination,~~ inspection, verification (including production process verification, use of statistical techniques for product acceptance, and related instructions for acceptance by the organization, and as applicable critical items including key characteristics,*

AS9100C moved the old AS9100B entries (h) and (j) to be the third and fourth entries under subclause (g):

> *g) requirements ~~relative~~ regarding the need for the supplier to*
>
> *- ~~supplier notification to~~ notify the organization of nonconforming product ~~and~~*
>
> *- ~~arrangements for~~ obtain organization approval ~~of supplier~~ for nonconforming ~~material,~~ product disposition,*
>
> *- ~~requirements for the supplier to~~ notify the organization of changes in product and/or process ~~definition~~, changes of suppliers, changes of manufacturing facility location and, where required, obtain organization approval, and*
>
> *- ~~requirements for the supplier to~~ flow down to ~~sub-tier suppliers~~ the supply chain the applicable requirements ~~in the purchasing documents,~~ including ~~key characteristics where required~~ customer requirements.*

AS9100C replaced the old subclause (h) with a new entry regarding record retention, and revised the existing subclause (i):

> *h) records retention requirements, and*
>
> *i) right of access by the organization, their customer, and regulatory authorities to the applicable areas of all facilities, at any level of the supply chain, involved in the order and to all applicable records~~, and~~.*

7.4.3 Verification of Purchased Product

AS9100C used the last paragraph of clause 7.4.3 to create a new Note:

> *NOTE 1: ~~Verification by the~~ Customer verification activities performed at any level of the supply chain should not be used by the organization or the supplier as evidence of effective control of quality ~~by the supplier~~ and ~~shall~~ does not absolve the organization of ~~the~~ its responsibility to provide acceptable product~~, nor shall it preclude subsequent rejection by the customer~~ and comply with all requirements.*

AS9100C moved requirements (a) through (e) to a second new Note. The only revised text appears in the first entry (old subclause [a]) under the note:

> *NOTE 2: Verification activities ~~may~~ can include*
>
> *~~a)~~ - obtaining objective evidence of the ~~quality~~ conformity of the product from the suppliers ~~(~~e.g., accompanying documentation, certificate of conformity, test reports, statistical records, process control records),*

AS9100C rewrote the requirement in clause 7.4.3 for a positive recall procedure as shown below:

> *Where purchased product is released for production use pending completion of all required verification activities, it shall be identified and recorded to allow recall and replacement if it is subsequently found that the product does not meet requirements.*
>
> *~~Purchased product shall not be used or processed until it has been verified as conforming to specified requirements unless it is released under positive recall procedure.~~*

AS9100C deleted the following requirement from the standard because it was viewed as too prescriptive and not applicable to all organization sizes and product types.

> *~~Where the organization utilizes test reports to verify purchased product, the data in those reports shall be acceptable per applicable specifications. The organization shall periodically validate test reports for raw material.~~*

AS9100C also deleted:

> *~~Where specified in the contract, the customer or the customer's representative shall be afforded the right to verify at the supplier's premises and the organization's premises that subcontracted product conforms to specified requirements.~~*

7.5 Production and Service Provision

7.5.1 Control of Production and Service Provision

AS9100C moved the opening text about planning, along with its four entries, to the end of clause 7.5.1. New Notes have been added under sub-clauses (a), (b), and (c):

> a) the availability of information that describes the characteristics of the product,
>
> *NOTE: This information can include drawings, parts lists, materials, and process specifications.*
>
> b) the availability of work instructions, as necessary,
>
> *NOTE: Work instructions can include process flow charts, production documents, (e.g., manufacturing plans, travelers, routers, work orders, process cards), and inspection documents.*
>
> c) the use of suitable equipment,
>
> *NOTE: Suitable equipment can include product specific tools (e.g., jigs, fixtures, molds) and software programs.*

Clause 7.5.1, subclauses (d) and (f) were modified by ISO 9001:2008. The title of clause 7.6 was changed to refer to the control of monitoring and measuring "equipment" instead of "devices," therefore, the terminology has been changed below:

> *d) the availability and use of monitoring and measuring ~~devices~~ equipment,*

The change below simply clarifies that implementation activities are those related to the "product."

> *f) the implementation of product release, delivery, and post-delivery activities.*

AS9100C revised subclauses (g), (h), (j), and (k):

> *g) accountability for all product during ~~manufacture~~ production (e.g., parts quantities, split orders, nonconforming product),*
>
> *h) evidence that all ~~manufacturing~~ production and inspection/verification operations have been completed as planned, or as otherwise documented and authorized,*
>
> *j) monitoring and control of utilities and supplies ~~such as~~ (e.g., water, compressed air, electricity, ~~and~~ chemical products) to the extent they affect conformity to product ~~quality~~ requirements, and*
>
> *k) criteria for workmanship, ~~which shall be stipulated~~ specified in the clearest practical manner (e.g., written standards, representative samples, ~~or~~ illustrations).*

AS9100C revised and moved this planning text from the top of the section to the end of the section:

> *Planning shall consider, as ~~applicable~~ appropriate*
>
> *- the establishment of process controls and development of control plans where key characteristics have been identified,*
>
> *- ~~the~~ designing, manufacturing, and using ~~of~~ tooling to measure variable data, ~~so that variable measurements can be taken, particularly for key characteristics, and~~*
>
> *- ~~the identification of~~ identifying in-process inspection/verification points when adequate verification of conformance cannot be performed at ~~a~~ later stages of realization,*
>
> *- special processes (see 7.5.2).*

AS9100C dropped the old section 7.5.1.1 on Production Documentation. A revised version of that old text, 7.5.1.1 (a), was moved under 7.5.1 (a) as a new Note. Likewise, a revised version of old clause 7.5.1.1 (b) was moved under 7.5.1 (b) as a new Note.

The new 7.5.1.1 is titled Production Process Verification and includes revised text from clause 8.2.4.2 regarding first article inspection (FAI). Placing the text in section 7 acknowledges that FAI is not primarily a monitoring and measuring process, but one that is used to assure product realization under controlled conditions. And, being in section 7, it can now be excluded for unique and individual products.

> *7.5.1.1 Production Process Verification*
>
> *The organization~~'s system~~ shall ~~provide~~ use ~~a process for the inspection, verification, and documentation of~~ a representative item from the first production run of a new part or assembly to verify that the production processes, production documentation, and tooling are capable of producing parts and assemblies that meet requirements. This process shall be repeated ~~or following~~ when ~~any subsequent~~ changes occur that invalidates the ~~previous~~ original ~~first article inspection~~ results (e.g., engineering changes, manufacturing process changes, tooling changes).*
>
> *NOTE: This activity is often referred to as first article inspection. ~~See (AS) (EN) (SJAC) 9102 for guidance.~~*

AS9100C retained the placement and title for clause 7.5.1.2. It has been revised to clarify that the referenced programs to be controlled and documented are "software" programs. Also, the changes made by ISO 9001:2008 elsewhere to refer to product "conformity" instead of product "quality" are likewise made in this section for consistency.

7.5.1.2 Control of Production Process Changes:

Personnel ~~Persons~~ authorized to approve changes to production processes shall be identified.

The organization shall identify and obtain acceptance of changes that require customer and/or regulatory authority approval in accordance with contract or regulatory requirements.

The organization shall control and document changes affecting processes, production equipment, tools, ~~and~~ or software programs ~~shall be documented~~. ~~Procedures shall be available to control their implementation.~~

The results of changes to production processes shall be assessed to confirm that the desired effect has been achieved without adverse effects to product ~~quality~~ conformity.

AS9100C changed the title for this clause to refer to the broader term, "software" programs, instead of using the more limited term, "numerical control" programs.

7.5.1.3 Control of Production Equipment, Tools, and ~~Numerical Control (NC) Machine~~ Software Programs:

Production equipment, tools and software programs used to automate and control/monitor product realization processes, shall be validated prior to ~~use~~ release for production and shall be maintained ~~and inspected periodically according to documented procedures. Validation prior to production use shall include verification of the first article produced to the design data/specification~~.

Storage requirements, including periodic preservation/condition checks, shall be ~~established~~ defined for production equipment or tooling in storage.

AS9100C revised and moved clause 7.5.1.4 to clause 7.1.4, with a new title, "Control of Work Transfers."

~~7.5.1.4 Control of Work Transferred, on a Temporary Basis, Outside the Organization's Facilities:~~

~~When planning to temporarily transfer work to a location outside the organization's facilities, the organization shall define the process to control and validate the quality of the work.~~

AS9100C changed the title of clause 7.5.1.4 to "Post-Delivery Support."

7.5.1.4 ~~Control of Service Operations~~ Post-Delivery Support:

~~Where servicing is a specified requirement, service operation processes shall provide for~~

Post-delivery support shall provide as applicable the

a~~) a method of collecting~~ collection and ~~analyzing~~ analysis of in-service data,

b) actions to be taken, including investigation and reporting, ~~where~~ when problems are ~~identified~~ detected after delivery, ~~including investigation, reporting activities, and actions on service information consistent with contractual and/or regulatory requirements,~~

c) ~~the~~ control and updating of technical documentation,

d) ~~the~~ approval, control, and use of repair schemes, and

e) ~~the~~ controls required for off-site work (e.g., organization's work undertaken at the customer's facilities).

7.5.2 Validation of Processes for Production and Service Provision

The revised text in this clause makes clear that any process output that can't be verified may result in deficiencies becoming known only after the product is in use or the service has been delivered.

> *The organization shall validate any processes for production and service provision where the resulting output cannot be verified by subsequent monitoring or measurement~~. This includes any processes where~~ and, as a consequence, deficiencies become apparent only after the product is in use or the service has been delivered.*

AS9100C slightly changed the note below:

> *NOTE: These processes are ~~frequently~~ often referred to as special processes.*

AS9100C dropped the AS additions under 7.5.2 (a) and (c):

> The organization shall establish arrangements for these processes including, as applicable
>
> a) defined criteria for review and approval of the processes,
>
> ~~- qualification and approval of special processes prior to use,~~
>
> c) use of specific methods and procedures,
>
> ~~- control of the significant operations and parameters of special processes in accordance with documented process specifications and changes thereto,~~

7.5.3 Identification and Traceability

This clause continues to state that, where appropriate, the organization must identify the product by suitable means "throughout product realization." The text below refers to inspection and test status of the product, and some organizations may have thought it only applied to the final product. The revision below clarifies that identifying the product monitoring and measurement status applies throughout product realization, from received product to final product, including in-process product.

> *The organization shall identify the product status with respect to monitoring and measurement requirements throughout product realization.*

By moving the "records" reference to the end of the sentence below, the meaning has expanded from recording the product identification to keeping any type of record associated with product traceability.

> *Where traceability is a requirement, the organization shall control ~~and record~~ the unique identification of the product and maintain records (see 4.2.4).*

AS9100C moved the section on traceability from a requirement to this new Note:

> *Note: Traceability requirements can include ~~According to the level of traceability required by contract, regulatory, or other established requirement, the organization's system shall provide for:~~*
>
> *a) identification to be maintained throughout the product life;*
>
> *b) the ability to trace all ~~the~~ products manufactured from the same batch of raw material, or from the same manufacturing batch, ~~to be traced, as well as~~ to the destination (e.g., delivery, scrap) ~~of all products of the same batch~~;*
>
> *c) for an assembly, the ability to ~~identity~~ trace ~~of~~ its components to the assembly and then to ~~those of~~ the next higher assembly ~~to be traced~~;*
>
> *d) for a ~~given~~ product, a sequential record of its production (manufacture, assembly, inspection/verification) to be ~~retrieved~~ retrievable.*

AS9100B included a reference to clause 4.3, Configuration Management, in the ISO 9001:2008 Note below.
Since AS9100C moved Configuration Management to clause 7.1.3, the reference was modified.

> NOTE: In some industry sectors, configuration management is a means by which identification and traceability are maintained *(see ~~4.3~~ 7.1.3).*

7.5.4 Customer Property

The change below reads better, but hasn't changed the requirement to report customer property issues to the customer and keep records.

> *If any customer property is lost, damaged, or otherwise found to be unsuitable for use, ~~this shall be reported~~ the organization shall report this to the customer and ~~records maintained~~ maintain records (see 4.2.4).*

The existing Note in 7.5.4 has been revised to include "personal data" as an example of customer property, broadening its applicability to more organizations, especially service organizations.

The AS9100B addition to this ISO 9001:2000–based Note was removed by AS9100C after ISO 9001:2008 made the highlighted change.

> *NOTE: Customer property can include intellectual property and personal data. ~~including customer furnished data used for design, production and/or inspection.~~*

7.5.5 Preservation of Product

If anyone was confused over the meaning of "conformity of product" in the old text, using "conformity to requirements" should be easier to understand in the new text.

> *The organization shall preserve the ~~conformity of~~ product during internal processing and delivery to the intended destination in order to maintain conformity to requirements.*

The current requirement that begins with, "This preservation shall include," doesn't give the flexibility to include, or not include, the identification, handling, packaging, storage, and protection of the product. The change below allows product preservation to be applied as applicable.

> *~~This~~ As applicable, preservation shall include identification, handling, packaging, storage, and protection. Preservation shall also apply to the constituent parts of a product.*

AS9100C revised the text below to add "statutory" to "regulatory" requirements for consistency with ISO 9001:2008 and its combined use of statutory and regulatory to address "legal" requirements. The requirement on accompanying documentation was deleted.

> *Preservation of product shall also include, where applicable in accordance with product specifications and~~/or~~ applicable statutory and regulatory requirements, provisions for:*
>
> *a) cleaning;*
>
> *b) prevention, detection, and removal of foreign objects;*
>
> *c) special handling for sensitive products;*
>
> *d) marking and labeling including safety warnings;*
>
> *e) shelf life control and stock rotation;*
>
> *f) special handling for hazardous materials.*

AS9100C moved the requirement below from clause 7.5.5 to clause 8.2.4 since its focus is to monitor the product to ensure that all the documents required to accompany the product are present at delivery.

> *~~The organization shall ensure that documents required by the contract/order to accompany the product are present at delivery and are protected against loss and deterioration.~~*

7.6 Control of Monitoring and Measuring ~~Devices~~ Equipment

The second clause title to change in ISO 9001:2008 is clause 7.6, where "devices" has been changed to "equipment." The term "equipment" was already used in several places in clause 7.6. The term "devices" has a broader scope and could include non-equipment types of tools. Equipment is the better choice for this calibration clause.

The changes to the clause below were to replace "devices" with "equipment" and to remove the reference to clause 7.2.1, Determination of Requirements Related to the Product.

> *The organization shall determine the monitoring and measurement to be undertaken and the monitoring and measuring ~~devices~~ equipment needed to provide evidence of conformity of product to determined requirements ~~(see 7.2.1)~~.*

AS9100C revised this text to refer to equipment instead of devices, and to include verification.

> *The organization shall maintain a register of these monitoring and measuring ~~devices~~ equipment, and define the process employed for their calibration/verification including details of equipment type, unique identification, location, frequency of checks, check method and acceptance criteria.*

AS9100C revised this Note to refer to equipment instead of devices.

> *NOTE: Monitoring and measuring ~~devices~~ equipment includes, but ~~are~~ is not limited to: test hardware, test software, automated test equipment (ATE) and plotters used to produce inspection data. It also includes personally owned and customer supplied equipment used to provide evidence of product conformity.*

AS9100C slightly revised the text below, but did not change the requirement:

> *The organization shall ensure that environmental conditions are suitable for the calibration~~s~~, inspection~~s~~, measurement~~s,~~ and ~~tests~~ testing being carried out.*

A minor change to 7.6 (a) is shown below. This requirement went from "calibrated or verified" to "calibrated or verified, or both," meaning a type of equipment might be calibrated and/or verified.

> *Where necessary to ensure valid results, measuring equipment shall:*
>
> *a) be calibrated or verified, or both, at specified intervals, or prior to use, against measurement standards traceable to international or national measurement standards; where no such standards exist, the basis used for calibration or verification shall be recorded (see 4.2.4);*

AS9100C moved the recall requirement from subclause (f) to be a stand-alone sentence below the list.

> *~~f) be recalled to a defined method when requiring calibration.~~*
>
> *The organization shall establish, implement, and maintain a process for the recall of monitoring and measuring equipment requiring calibration or verification.*

The statement below, that measuring equipment must "be identified" sounded like the organization was to add identification. However, measuring equipment may come with the identification already in place.

> *c) ~~be identified~~ have identification ~~to enable~~ in order to determine ~~the~~ its calibration status ~~to be determined~~;*

The text below was split from its old paragraph and made a stand-alone sentence for emphasis.

> *Records of the results of calibration and verification shall be maintained (see 4.2.4).*

Software development organizations may have been unsure how to "confirm," per clause 7.6, that software used for monitoring and measurement has the ability to satisfy the intended application.

A new Note was added to explain that confirmation of software would typically include verification and configuration management.

> *NOTE: Confirmation of the ability of computer software to satisfy the intended application would typically include its verification and configuration management to maintain its suitability for use.*

This old Note for clause 7.6 was dropped. It referred the reader to the ISO 10012-1 and ISO 10012-2 standards for guidance. Although these standards have been replaced with ISO 10012:2003, the reference was not retained.

> *~~NOTE: See ISO 10012-1 and ISO 10012-2 for guidance.~~*

SECTION 8: MEASUREMENT, ANALYSIS, AND IMPROVEMENT

8.1—General
8.2—Monitoring and Measurement
8.3—Control of Nonconforming Product
8.4—Analysis of Data
8.5—Improvement

8.0 Measurement, Analysis, and Improvement

8.1 General

AS9100 The organization is required to plan for, establish, and improve processes to monitor, measure, and analyze its QMS and procedures, which must provide evidence that its products meet customer and other requirements, verify the QMS's effectiveness, and use and achieve continual improvement of the QMS. This will require identification of what statistical and other techniques are needed and how much they need to be used. ***Notes suggest design verification and process control.***

ISO 9004 Provides 11 topics to be evaluated in establishing measurement, analysis, and improvement processes. Offers guidance on applying measurement to improve the organization's performance into the future and to assess the value of the measurements in meeting its needs.

ISO 9000 Explores in 2.10, Role of Statistical Techniques, their use in understanding variability to help "improve effectiveness and efficiency."

Document Requirements:

None

Internal Audit Questions:

- Is there objective evidence to demonstrate that the organization has defined, planned, and implemented the measurement and monitoring activities needed to assure conformity and to achieve improvement?
- Is there objective evidence to demonstrate that the organization has determined the need for and use of applicable methodologies including statistical techniques (specifications, procedures, work instructions, control plans, process sheets, and so on)?

Management Summary:

Results of data analysis from improvement activities should be a management review input.

Management should note that measurements of customer satisfaction are vital for evaluating the organization's performance.

Process Control Tip:

Use analysis tools described in the Note of this AS9100 section to determine the best product characteristics or process steps in which to use statistical or other controls.

8.1 General

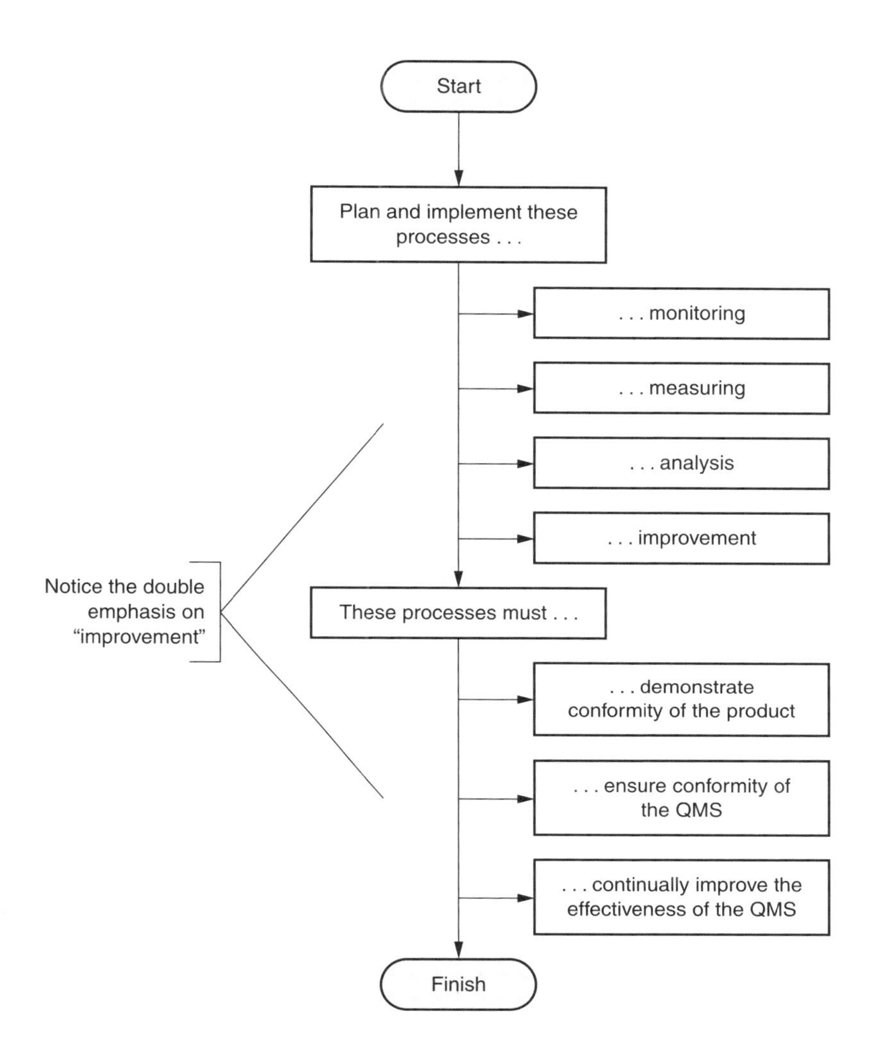
Start
Plan and implement these processes . . .
. . . monitoring
. . . measuring
. . . analysis
. . . improvement
Notice the double emphasis on "improvement"
These processes must . . .
. . . demonstrate conformity of the product
. . . ensure conformity of the QMS
. . . continually improve the effectiveness of the QMS
Finish

8.0 Measurement, Analysis, and Improvement

8.2 Monitoring and Measurement

8.2.1 Customer Satisfaction

AS9100 The organization must establish a program to collect, measure, and take action on data regarding customer satisfaction.

ISO 9004 Provides guidance on developing a system to measure customer satisfaction, including examples of available information relating to customers and their interaction with an organization, and of sources of information on customer satisfaction.

ISO 9000 Defines "customer satisfaction" in 3.1.4 as the customer's perception of how well its "requirements have been fulfilled."

Document Requirements:

None

Internal Audit Questions:

- Are customer satisfaction and/or customer dissatisfaction information surveyed, monitored, and acted on?
- Are methods for gathering and using customer information determined and deployed throughout the organization?
- Has the customer information been processed in such a way as to provide quantifiable data and trends?
- Are measures in place to ensure that requirements are met?

Management Summary:

The organization should use customer feedback to plan *processes* that listen effectively and efficiently to the "voice of the customer."

The organization should identify sources of customer and end user information (from internal and external sources) such as:

- Customer complaints
- Direct communication with customers
- Questionnaires and surveys
- Focus groups
- Consumer reports
- Various media reports
- Sector and industry studies

Process Control Tip:

Ensure customer satisfaction by creating processes that address customer requirements and are robust and repeatable.

8.2 Monitoring and Measurement

8.2.1 Customer Satisfaction

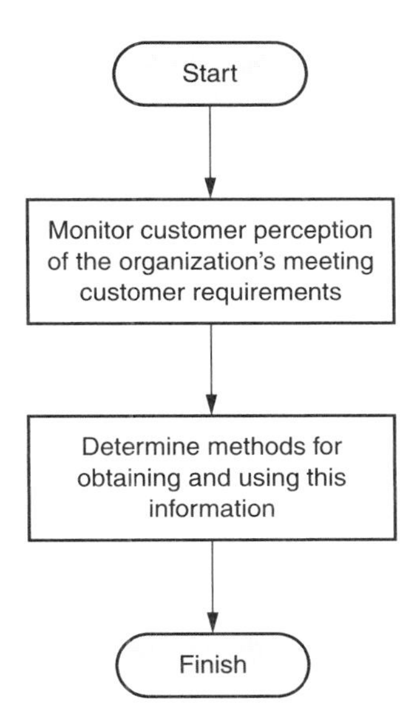

8.0 Measurement, Analysis, and Improvement

8.2 Monitoring and Measurement

8.2.2 Internal Audit

AS9100 A documented internal audit program must be established to assess the QMS's effectiveness in meeting the requirements of ISO 9001 and any other requirements the organization has specified for its QMS, and whether its QMS is functioning and being conformed to by employees in the course of organizational activities. The organization must plan out and conduct audits of each process, procedure, and area frequently and thoroughly enough to ensure the system's effective operation. Auditors must be capable of accurately assessing the areas they audit. Management for an area where a nonconformity is identified is responsible for ensuring that prompt and effective corrective action is taken, with the internal auditors responsible for verifying that the action eliminates the nonconformity and its causes. ***The standard recommends techniques to assure an effective internal auditing system that drives continuous improvement of the QMS, and meets contract or regulatory requirements.***

ISO 9004 Provides guidance on the purpose and management of internal audits and emphasizes the need for top management to take action to improve the QMS in response to internal audit findings. Offers 11 examples of topics to be examined in internal audits.

ISO 9000 Defines 14 terms in 3.9, Terms Relating to Audit, although it is noted that most of these terms and definitions are subject to change when ISO 19011, which will provide a single QMS and EMS auditing guidelines standard, is published. Explores role and types of audits in 2.8.2, Auditing the Quality Management System.

Document Requirements:

Documented procedure

Record

Internal Audit Questions:	*Management Summary:*
• Does the organization conduct periodic audits of the quality management system? • Do the periodic audits evaluate the conformity of the quality management system to the requirements of ISO 9001:2008? • Do the periodic audits evaluate the degree to which the quality management system has been effectively implemented and maintained? • When planning the audit program, does the organization consider the status and importance of areas to be audited, and results of previous audits? • Are the methodologies, audit scope, and frequency defined? • Do personnel other than those who perform the activity being audited perform the audits? • Is there a documented procedure that includes the responsibilities and requirements for conducting audits? • Is there a documented procedure that describes how to ensure the independence of auditors? • Is there a documented procedure for recording results and reporting to management? • Does management take timely corrective action on deficiencies found during the audit? • Do follow-up actions include the verification of the implementation of corrective action? • Do follow-up actions include the reporting of verification results?	The internal audit process is a management tool for independent assessment of a process or activity. Planning for internal audits should be flexible in order to allow changes in emphasis based on findings and objective evidence found during the audit. Management should provide opportunities for recognition of areas that achieve excellent performance during an internal audit.

8.2 Monitoring and Measurement

8.2.2 Internal Audit

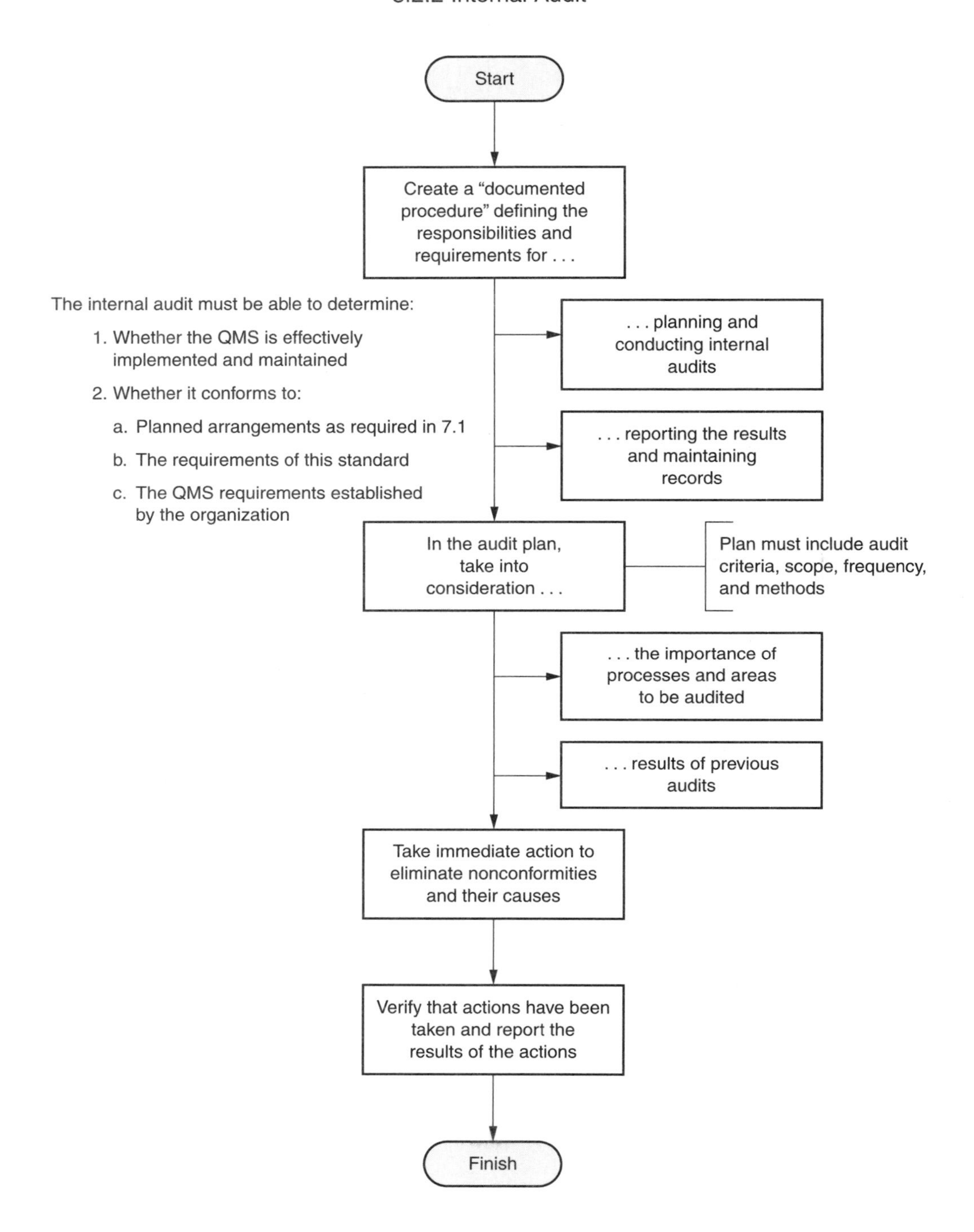

8.0 Measurement, Analysis, and Improvement

8.2 Monitoring and Measurement

8.2.3 Monitoring and Measurement of Processes

AS9100 QMS processes must be monitored and measured to the degree necessary to ensure that the processes produce conforming product and achieve expected outcomes, with corrective actions to be taken when the processes do not produce expected outcomes. ***Variance from the standard process may create product nonconformance and shall be corrected. Processes address identification and control of nonconforming product per clause 8.3.***

ISO 9004 Provides guidance on the use and advantages of self-assessments (including Annex A, Guidelines for Self-Assessment) to improve performance and on the use of QMS information to measure financial performance. Recommends measuring process performance and includes examples of performance aspects that can be measured.

ISO 9000 Explores in 2.8.1, Evaluating Processes within the Quality Management System, the questions to be addressed in evaluating a QMS.

Document Requirements:

None

Internal Audit Questions:

- Has the organization identified the key realization processes necessary to meet customer requirements?
- Has the organization employed suitable methods to measure and monitor key realization processes?
- Are the intended purposes of the key realization processes quantified by process parameter specifications, by specifications for the product output of the process, or by some other means?
- Are the measurement and monitoring methods for realization processes adequate for confirming the continuing suitability of each process to satisfy its intended purpose?

Management Summary:

The organization should incorporate measurements of process performance into processes and use the measurements in process management.

Measurements of process performance include:

- Cycle time or throughput
- Dependability
- Yield
- Technology utilization
- Waste reduction
- Cost allocation and reduction

Process Control Tip:

Documented, visual, and physical process controls, as mentioned throughout this guide, assure the fulfillment of the requirements of this element of the standard.

8.2 Monitoring and Measurement

8.2.3 Monitoring and Measurement of Processes

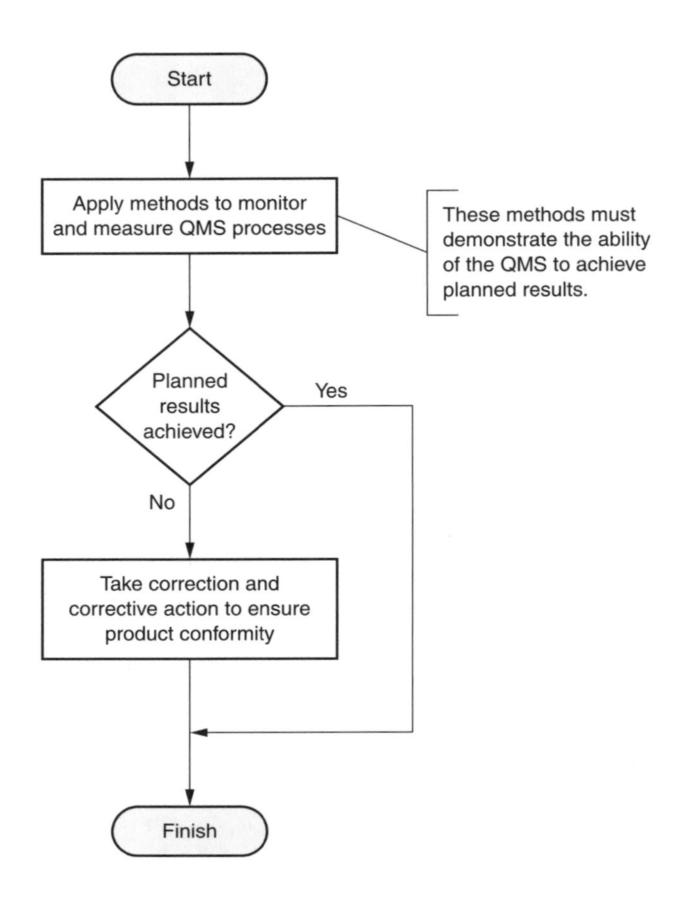

8.0 Measurement, Analysis, and Improvement

8.2 Monitoring and Measurement

8.2.4 Monitoring and Measurement of Product

AS9100 The organization must engage in monitoring and measurement activities throughout its production processes to verify that product conforms to customer and other requirements, "maintain" its evidence of conformance, and treat documentation identifying who is responsible for a product's release as quality records. A product must not be released until monitoring and measurement indicate that the product conforms to requirements unless the employee responsible for its release authorizes the release beforehand—and the action is acceptable to the customer. The same requirements apply to service delivery. ***Monitoring and controlling of key characteristics shall be planned, and product acceptance measures shall assure conformance of product, be statistically valid, and accepted by the customer as required.***

Inspection documentation: Document measurement requirements, criteria, stages, and instructions for product and service acceptance. Assure that product qualification records record actual data where required and fully demonstrate that product meets specifications.

First article inspection: First runs or changes to previously verified processes that produce product require representative items that confirm the process produced conforming product.

ISO 9004 Recommends using product measurement to improve production processes. Offers 10 aspects and elements of the measurement process that should be considered in selecting measurement methods, and four examples of measurement records to be kept.

ISO 9000 Defines "release" in 3.6.9 as "permission to proceed to the next stage of a process."

Document Requirements:

Record

Internal Audit Questions:

- Does the organization measure and monitor product characteristics to verify that product requirements are met?
- Does the organization measure and monitor product characteristics at appropriate stages of the product realization process?
- Is there objective evidence that acceptance criteria for product have been met?
- Is there identification of the authority responsible for release of the product?
- Are all specified activities performed prior to product release and service delivery?
- When specified activities have not been performed prior to product release or service delivery, is the customer informed?
- Has the customer approved of the action?

Management Summary:

The organization should review the methods used for measuring products and verification records, to improve performance.

Process Control Tip:

Control plans list measurements and serve as documentation that the requirements are conveyed to those working the process of product acceptance.

Quality planning and contract review processes assure that control plans define measurement plans and can incorporate first article inspection requirements. Assure that quality planning follows the provisions of AS9102.

RTA documentation can assure that acceptance criteria are best adhered to up front, during software design.

8.2 Monitoring and Measurement

8.2.4 Monitoring and Measurement of Product

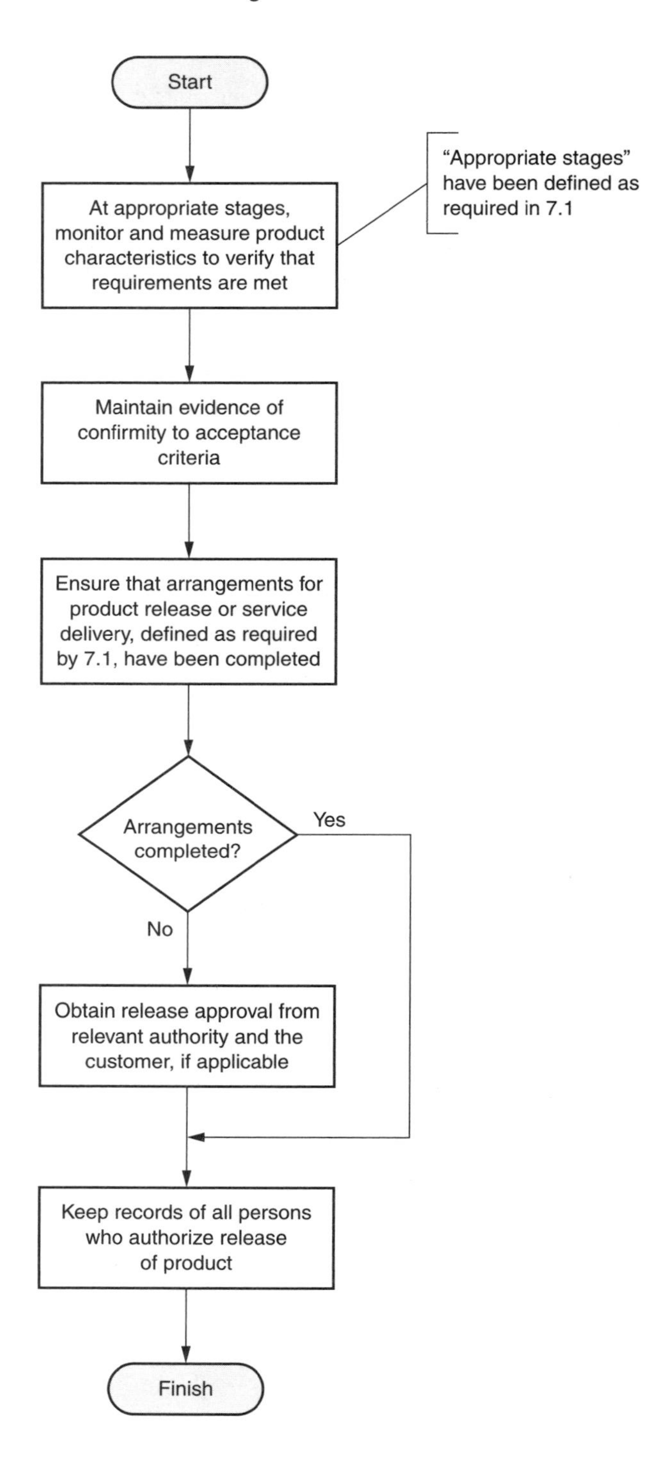

8.0 Measurement, Analysis, and Improvement

8.3 Control of Nonconforming Product

AS9100 The organization must prevent unintended delivery or use of nonconforming product, with documentation of the processes to be followed and the authorities responsible for identifying, controlling, and disposing of nonconforming product. The organization must correct nonconformity, offer the product on an "as is" basis, or scrap it, with information on the nonconformity and the product's disposition to be treated as a quality record. Corrected product must be reevaluated to ensure that the nonconformity has been eliminated, and the organization is responsible for resolving known or potential nonconformities that arise after a product is delivered or in use, ***including customer returned product. Assure that procedures address review and authority for specific dispositions as authorized by the customer, including permanently marking or rendering scrap product unusable. The organization must notify the customer, regulatory agency, or any other applicable outside party upon discovery of any nonconforming product that may affect reliability and safety.***

ISO 9004 Examines how top management can establish procedures to deal with nonconforming product that will not only prevent unintended use, but will provide feedback to the organization for quality planning and process improvement.

ISO 9000 Defines terms relating to the disposition of nonconforming product in 3.6.8, Release, 3.6.10, Repair, 3.6.11, Rework, 3.6.12, Re-grade, and 3.6.13, Scrap.

Document Requirements:

Documented procedure

Record

Internal Audit Questions:

- Is there a documented procedure to assure that product that does not conform to requirements is identified and controlled to prevent unintended use or delivery?
- Is there evidence of appropriate action being taken when nonconforming product has been detected after delivery or use has started?
- Is it required that any proposed rectification of nonconforming product be reported for concession to the customer, the end user, or a regulatory body?
- Is there objective evidence of appropriate communication with a customer when the organization proposes rectification of nonconforming product?

Management Summary:

To provide analysis data and improvement activities, the organization should record nonconformities to both product realization and support processes.

During reviews of nonconformities, negative trends should be considered for improvement as well as input to management review for consideration.

Acceptance of nonconformity disposition may be a contractual requirement of the customer.

Process Control Tip:

Use lean principles to easily identify and control nonconforming material.

8.3 Control of Nonconforming Product

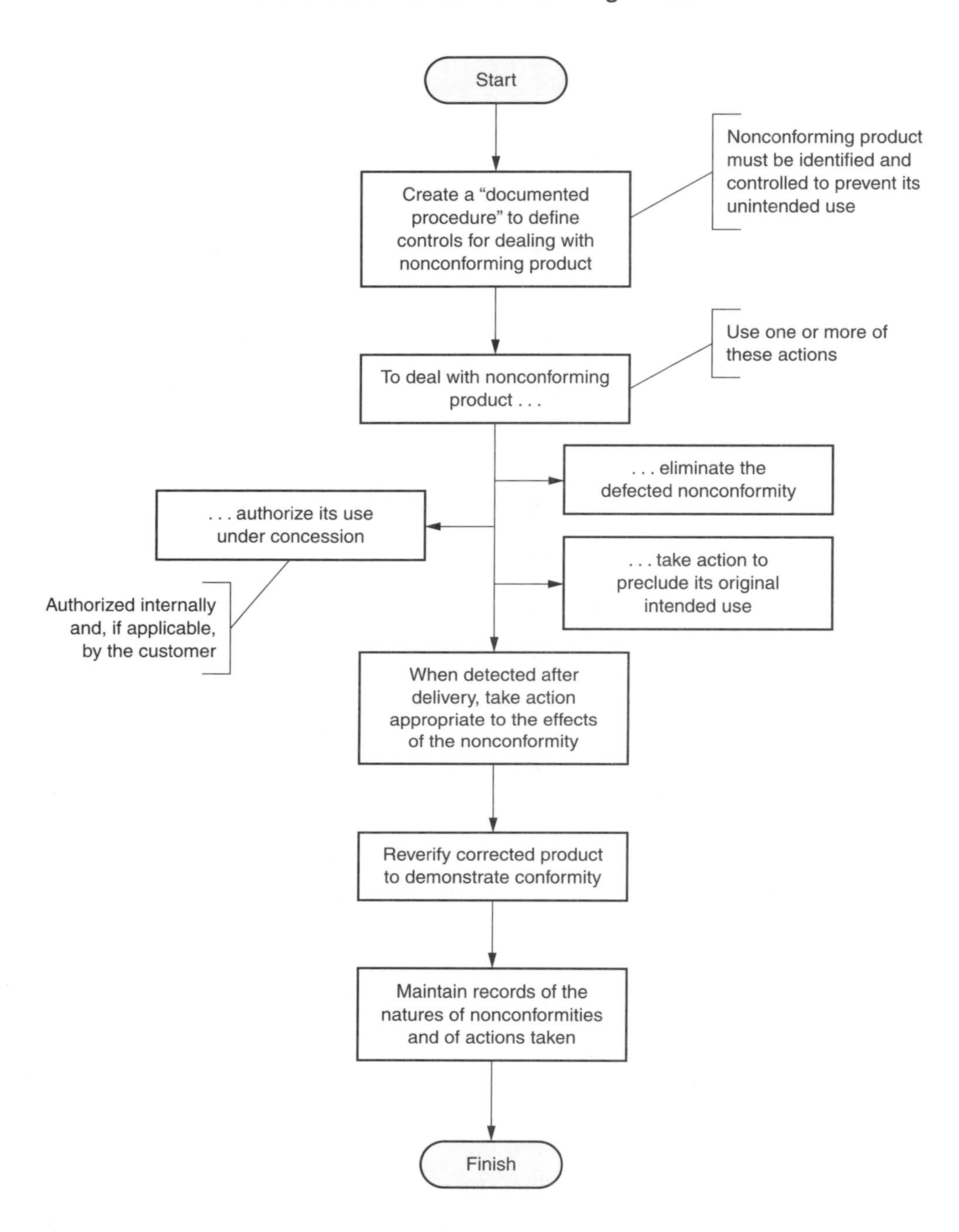

8.0 Measurement, Analysis, and Improvement

8.4 Analysis of Data

AS9100 The organization must use data from its monitoring and measuring activities and other sources to verify that the QMS is appropriate to the organization's needs and effectively conforms to ISO 9001 and other requirements, and to identify opportunities for improvement of the QMS. It requires the resulting analysis to provide data on four subjects relating to the QMS and the organization's products.

ISO 9004 Stresses the value of fact-based measurements and of identifying root causes of problems. Identifies nine purposes to which analysis can be put.

Document Requirements:

None

Internal Audit Questions:

- Has the organization determined the appropriate data to be collected to determine the effectiveness of the quality management system and where improvements can be made?
- Does the organization analyze appropriate data to determine the suitability and effectiveness of the quality management system and improvements that can be made?
- Does the organization analyze appropriate data to provide information on customer satisfaction and/ or dissatisfaction and conformance to customer requirements?
- Does the organization analyze appropriate data to provide information on characteristics of processes, product, and their trends?
- Does the organization analyze appropriate data to provide information on suppliers?

Management Summary:

The organization should analyze data to assess performance against plans, objectives, and defined goals and to identify areas for improvement.

8.4 Analysis of Data

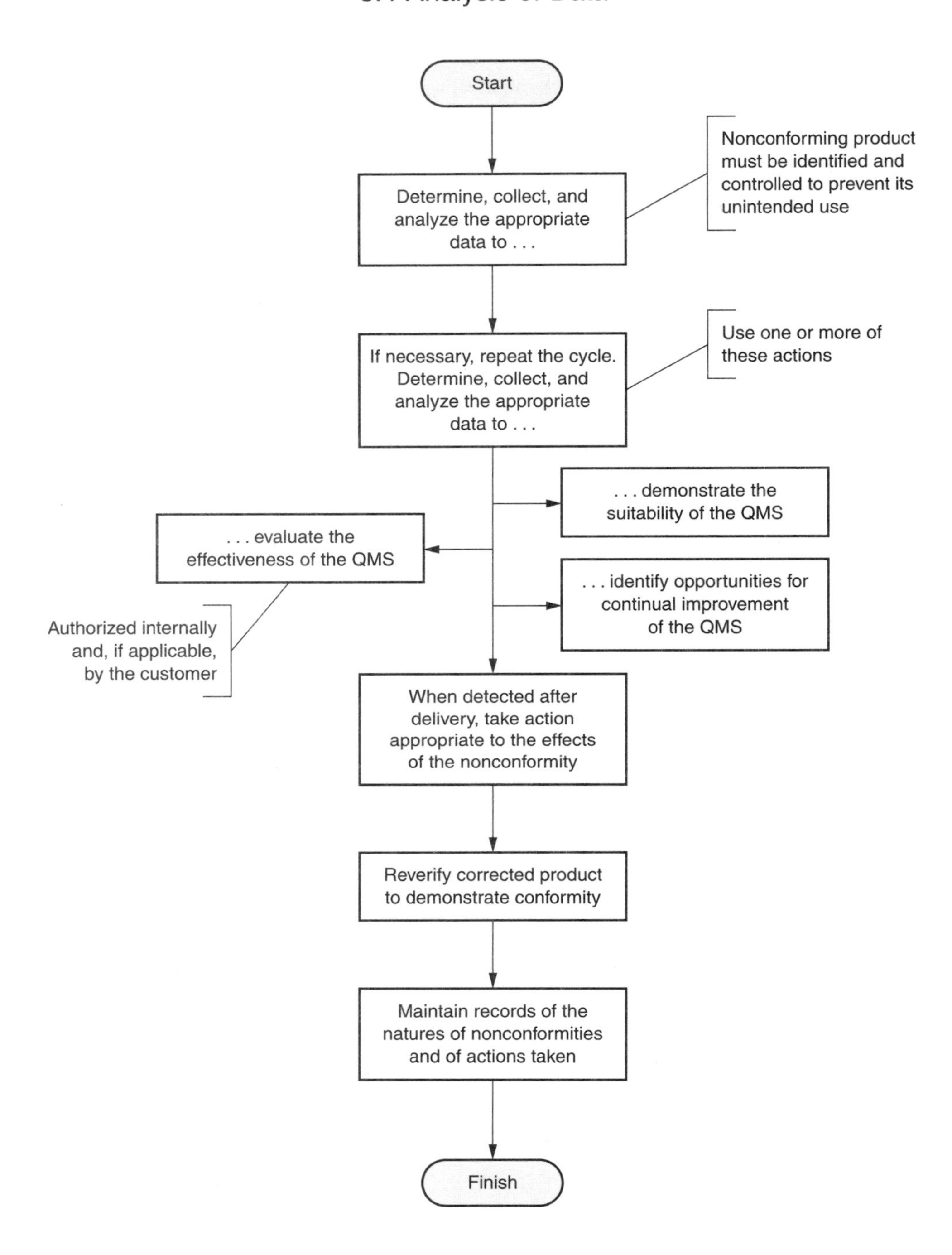

8.0 Measurement, Analysis, and Improvement

8.5 Improvement

8.5.1 Continual Improvement

AS9100	The organization is required to use seven specified QMS elements/processes, all required by ISO 9001, to "continually improve the effectiveness" of the QMS.
ISO 9004	Explores the range of improvements management can pursue and encourages a proactive approach.
ISO 9000	Defines "continual improvement" in 3.1.13 as "recurring activity to increase the ability to fulfill requirements" and explores in 2.9, Continual Improvement, the purpose of and actions that lead to continual improvement.

Document Requirements:

None

Internal Audit Questions:

- Does the organization plan and manage processes necessary for continual improvement of the quality management system?
- Does the organization use the quality policy, quality objectives, and analysis of data to facilitate continual improvement of the quality management system?
- Does the organization use audit results, corrective action, and preventive action to facilitate continual improvement of the quality management system?
- Does the review of quality data result in action items designed to improve the quality management system (also see section 5.6.3)?

Management Summary:

Top management should define and implement a process for continual improvement that can be applied to realization and support processes.

Top management should create an environment that challenges people to seek opportunities for improvement of performance in processes, activities, and products.

8.5 Improvement

8.5.1 Continual Improvement

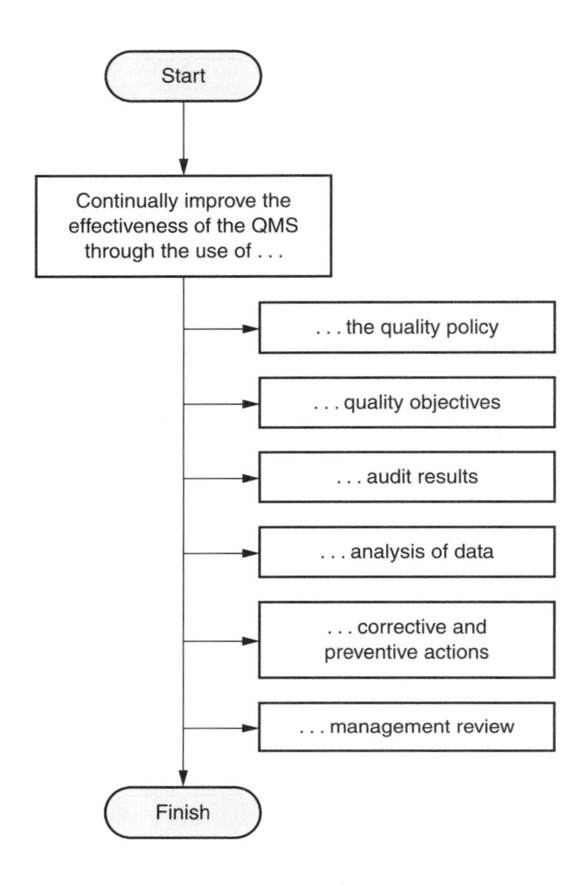

8.0 Measurement, Analysis, and Improvement

8.5 Improvement

8.5.2 Corrective Action

AS9100	The organization must document the processes to be used to assess and correct nonconformities, eliminate their cause(s) to prevent future nonconformities from the same cause, and evaluate the effectiveness of the correction. Corrective action must be proportionate to the severity of the nonconformity, ***directed to any party internal or external to the organization that may have generated the root cause, and timely and effective, with*** documentation of corrective actions to be treated as a quality record.
ISO 9004	Provides guidance on planning for corrective actions and determining causes. Offers 10 examples of information resources that help define effective corrective actions.
ISO 9000	Defines "corrective action" in 3.6.5 as "action taken to eliminate the cause of a detected nonconformity or other undesirable situation" and notes that it is meant to prevent recurrence of nonconformity, while "preventive action" is meant to avoid the possibility of a nonconformity ever occurring.

Document Requirements:

Documented procedure

Record (e)

Internal Audit Questions:

- Does the organization take corrective action to eliminate causes of nonconformities?
- Is corrective action taken appropriate to the impact of the problems encountered?
- Do corrective action procedures provide for identifying nonconformities, determining causes, evaluating need for actions to prevent recurrence, determining the corrective action needed, and implementation of the needed corrective action?
- Do corrective action procedures provide for recording the results of corrective actions taken?
- Do the preventive action documented procedures provide for reviewing the corrective action taken?

Management Summary:

Corrective action should be used as a tool for improvement.

Corrective actions should be included in management review.

The organization should incorporate root cause analysis into the corrective action process.

Corrective action considerations include:

- Customer complaints
- Nonconformity reports
- Internal audit reports
- Management review outputs

8.5 Improvement

8.5.2 Corrective Action

Corrective actions eliminate the causes of nonconformities to prevent recurrence

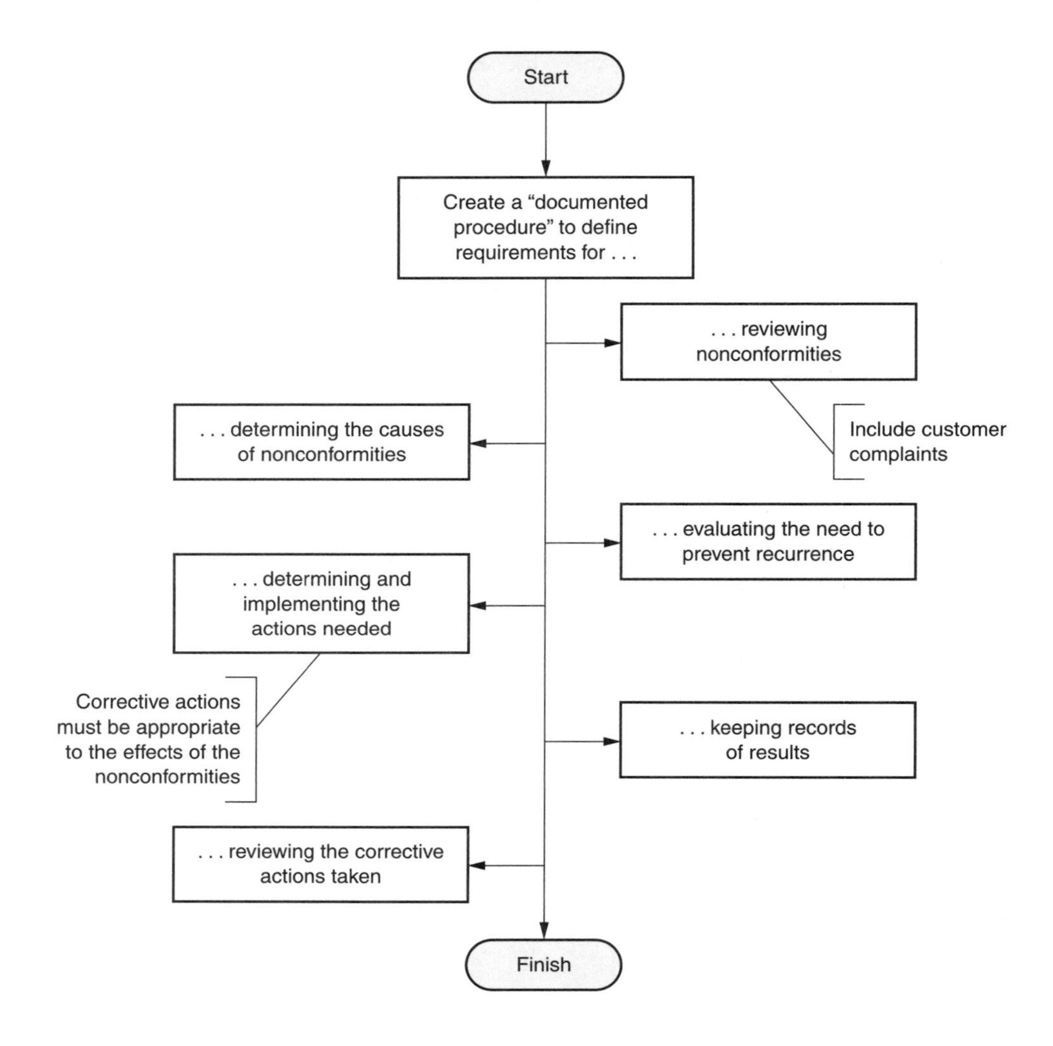

8.0 Measurement, Analysis, and Improvement

8.5 Improvement

8.5.3 Preventive Action

ISO 9001 The organization must document the processes to be used to prevent nonconformities from ever occurring by assessing and eliminating their potential causes—with action proportionate to the severity of a potential nonconformity—and evaluating the effectiveness of the actions. Documentation of preventive actions is to be treated as quality records.

ISO 9004 Recommends "planning for loss prevention" and offers 12 sources for quantitative data to be used in the planning process.

Document Requirements:

Documented procedure

Record (d)

Internal Audit Questions:

- Does the organization identify preventive actions needed to eliminate the potential causes of possible nonconformities?
- Is preventive action taken appropriate to the impact of potential problems?
- Do the preventive action documented procedures provide for identifying potential nonconformities and their probable causes?
- Do the documented procedures for preventive action provide for determining the need for preventive action and implementation of the preventive action needed?
- Do the preventive action documented procedures provide for recording the results of the preventive actions taken?
- Do the documented procedures for preventive action provide for reviewing the preventive action taken?

Management Summary:

The organization should develop a plan for loss prevention and apply it to realization and support processes, activities, and products.

Results of the evaluation of the loss prevention plan should be considered as an output for management review and should be used in the improvement processes.

8.5 Improvement

8.5.3 Preventive Action

Preventive actions eliminate the causes of nonconformities before they occur

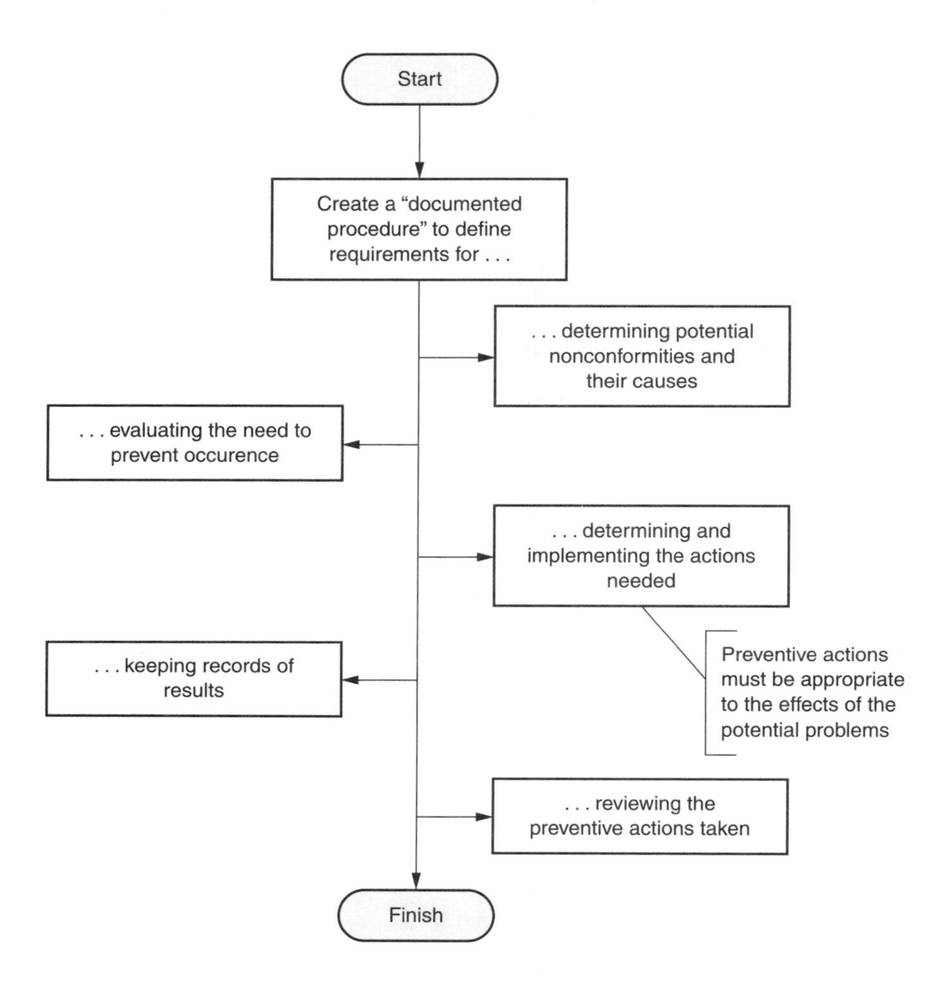

REVISIONS TO AS9100:2009 (REVISION C), CLAUSE 8

(Source: Larry Whittington)

8. Measurement, Analysis, and Improvement

8.1 General

The prior use of "conformity of the product" has been revised to "conformity to product requirements."

> *The organization shall plan and implement the monitoring, measurement, analysis, and improvement processes needed*
>
> *a) to demonstrate conformity ~~of the~~ to product requirements,*

AS9100C modified the Note below to switch from "may" to "can," indented four of the entries under "process control," and added "criticality" to the last entry regarding analysis.

> *NOTE: According to the nature of the product and depending on the specified requirements, statistical techniques ~~may~~ can be used to support:*
>
> *- design verification (e.g., reliability, maintainability, safety),*
>
> *- process control,*
>
> *- selection and inspection of key characteristics,*
>
> *- process capability measurements,*
>
> *- statistical process control,*
>
> *- design of experiment,*
>
> *- inspection, and - ~~matching sampling rate to the criticality of the product and to the process capability~~;*
>
> *- failure mode, ~~and~~ effect and criticality analysis.*

8.2 Monitoring

8.2.1 Customer Satisfaction

AS9100C expanded on the customer satisfaction requirement to include specific types of measurements and to require that plans be implemented to improve the customer's satisfaction. This change establishes a clear link between the quality management system and organizational performance.

> *Information to be monitored and used for the evaluation of customer satisfaction shall include, but is not limited to, product conformity, on-time delivery performance, customer complaints, and corrective action requests. Organizations shall develop and implement plans for customer satisfaction improvement that address deficiencies identified by these evaluations, and assess the effectiveness of the results.*

A new Note has been added to clause 8.2.1 to provide examples of monitoring customer perceptions.

> *NOTE: Monitoring customer perception can include obtaining input from sources such as customer satisfaction surveys, customer data on delivered product quality, user opinion surveys, lost business analysis, compliments, warranty claims, dealer reports.*

8.2.2 Internal Audit

AS9100C added a Note under clause 8.2.2 (a) to clarify that "planned arrangements" includes customer contractual requirements:

> a) conforms to the planned arrangements (see 7.1), to the requirements of this International Standard, and to the quality management system requirements established by the organization, and
>
> *NOTE: Planned arrangements include customer contractual requirements.*

The change below was to simply add the word "The" at the beginning of the sentence.

> *The ~~S~~selection of auditors and conduct of audits shall ensure objectivity and impartiality of the audit process.*

The requirement below was edited to emphasize the need for a documented procedure (by placing it first in the sentence). Also, "establishing records" has been moved ahead of "reporting results" in the list of topics to be defined in the procedure. Records are being captured throughout the audit and should be listed before the reporting of results.

> *A documented procedure shall be established to define the responsibilities and requirements for planning and conducting audits, establishing records, and reporting results.*
>
> ~~*The responsibilities and requirements for planning and conducting audits, and for reporting results and maintaining records (see 4.2.4) shall be defined in a documented procedure.*~~

The new sentence below highlights the need to maintain records of the audit and its results. The reference in the old text to 4.2.4 for record control was moved to this new sentence.

> *Records of the audits and their results shall be maintained (see 4.2.4).*

Expanding from "actions" to "any necessary corrections and corrective actions" reminds us that an immediate correction might be needed before determining the cause of the nonconformity and taking corrective action to prevent its recurrence. Clause 8.2.3 also refers to corrections and corrective actions.

> *The management responsible for the area being audited shall ensure that any necessary corrections and corrective actions are taken without undue delay to eliminate detected nonconformities and their causes.*

The reference in the Note below to the withdrawn ISO 10011 standard has been replaced with a reference to ISO 19011, *Guidelines for quality and/or environmental management systems auditing.*

> *NOTE: See ~~ISO 10011-1, ISO 10011-2 and ISO 10011-3~~ ISO 19011 for guidance.*

AS9100C deleted the following text from clause 8.2.2 because it was viewed as too prescriptive:

> ~~*Detailed tools and techniques shall be developed such as checksheets, process flowcharts, or any similar method to support audit of the quality management system requirements. The acceptability of the selected tools will be measured against the effectiveness of the internal audit process and overall organization performance.*~~

AS9100C could delete this text because a general statement to this effect was added to clause 4.1 for application across the entire quality management system. Plus, a new Note was added earlier in the section about internal audit planned arrangements including customer contractual requirements.

> ~~*Internal audits shall also meet contract and/or regulatory requirements.*~~

8.2.3 Monitoring and Measurement of Processes

This clause requires applying suitable methods for monitoring and measuring processes to demonstrate their ability to achieve planned results. For some supporting processes, these results are only indirectly related to product conformity. Therefore, the reference to product conformity was moved from this paragraph to the new Note below.

> *When planned results are not achieved, correction and corrective action shall be taken, as appropriate*~~*, to ensure conformity of the product*~~*.*

What is a "suitable" method for monitoring and measuring processes? The Note below says to consider the type and extent of monitoring or measurement based on the impact of the process on conformity to product requirements and system effectiveness.

> *NOTE: When determining suitable methods, it is advisable that the organization consider the type and extent of monitoring or measurement appropriate to each of its processes in relation to their impact on the conformity to product requirements and on the effectiveness of the quality management system.*

AS9100C added a new subclause to the activities performed in the event of a process nonconformity:

> *In the event of process nonconformity, the organization shall*
>
> *a) take appropriate action to correct the nonconforming process,*
>
> *b) evaluate whether the process nonconformity has resulted in product nonconformity,* ~~*and*~~
>
> *c) determine if the process nonconformity is limited to a specific case or whether it could have affected other processes or products, and*
>
> *d) identify and control the nonconforming product* ~~*in accordance with clause*~~ *(see 8.3).*

8.2.4 Monitoring and Measurement of Product

> *The organization shall monitor and measure the characteristics of the product to verify that product requirements have been met. This shall be carried out at appropriate stages of the product realization process in accordance with the planned arrangements (see 7.1). Evidence of conformity with the acceptance criteria shall be maintained.*

The requirement to maintain evidence of conformity with acceptance criteria has been moved from the paragraph below to the first paragraph above. And the release of product is not to the next in-process stage, but for delivery to the customer.

> *~~Evidence of conformity with the acceptance criteria shall be maintained~~. Records shall indicate the person(s) authorizing release of product for delivery to the customer (see 4.2.4).*

AS9100C removed clause 8.2.4.1 and left most of that section as an embedded part of clause 8.2.4:

> *~~8.2.4.1 Inspection Documentation:~~*
>
> *Measurement requirements for product ~~or service~~ acceptance shall be documented~~. This documentation may be part of the production documentation, but~~ and shall include*
>
> *a) criteria for acceptance and/or rejection,*
>
> *b) where in the sequence measurement and testing operations are performed,*
>
> *c) ~~a~~ required records of the measurement results (at a minimum, indication of acceptance or rejection), and*
>
> *d) ~~type of~~ any specific measurement instruments required and any specific instructions associated with their use.*
>
> *~~Test records shall show actual test results data when required by specification or acceptance test plan.~~*
>
> *~~Where required to demonstrate product qualification the organization shall ensure that records provide evidence that the product meets the defined requirements.~~*

AS9100C expanded this section to apply to critical items, not just key characteristics:

> *When critical items, including key characteristics, have been identified the organization shall ensure they ~~shall be~~ are controlled and monitored ~~and controlled~~ in accordance with the established processes.*

AS9100C clarified this section on sampling inspection and the appropriateness of the sampling plan:

> *When the organization uses sampling inspection as a means of product acceptance, the plan shall be ~~statistically valid~~ justified on the basis of recognized statistical principles and appropriate for use (i.e., matching the sampling plan to the criticality of the product and to the process capability) ~~. The plan shall preclude the acceptance of lots whose samples have known nonconformities. When required, the plan shall be submitted for customer approval.~~*

AS9100C revised this section on the positive recall process:

> *~~Product shall not be used until it has been inspected or otherwise verified as conforming to specified requirements, except when product is released under positive-recall procedures~~ Where product is released for production use pending completion of all required measurement and monitoring activities, it shall be identified and recorded to allow recall and replacement if it is subsequently found that the product does not meet requirements.*

This requirement was edited to clarify that the release of product and delivery of service is to the customer.

> *The release of product ~~release~~ and delivery of service ~~delivery~~ to the customer shall not proceed until the planned arrangements (see 7.1) have been satisfactorily completed, unless otherwise approved by a relevant authority and, where applicable, by the customer.*

AS9100C moved this requirement from clause 7.5.5 to clause 8.2.4 since it is more a product monitoring issue than one of product preservation.

> *The organization shall ensure that all documents required ~~by the contract/order~~ to accompany the product are present at delivery ~~and are protected against loss and deterioration~~.*

8.3 Control of Nonconforming Product

The sentence below has been edited to begin with (instead of end with) the requirement for a documented procedure.

> *A documented procedure shall be established to define ~~T~~the controls and related responsibilities and authorities for dealing with nonconforming product. ~~shall be defined in a documented procedure.~~*

AS9100C made a slight change to this existing Note:

> *NOTE: The term "nonconforming product" includes nonconforming product returned ~~from~~ by a customer.*

AS9100C changed this text to define both responsibility and authority, for both review and disposition:

> *The organization's documented procedure shall define the responsibility ~~for review~~ and authority for the review and disposition of nonconforming product and the process for approving personnel making these decisions.*

The requirement below adds "where applicable," meaning where relevant and suitable, to deal with nonconforming product in one or more of the ways listed.

> *Where applicable, ~~T~~the organization shall deal with nonconforming product by one or more of the following ways:*

The new entry below is edited text from the last sentence in clause 8.3 that has been moved to become part of the list of ways to deal with nonconforming product.

> *d) by taking action appropriate to the effects, or potential effects, of the nonconformity when nonconforming product is detected after delivery or use has started.*

AS9100C moved this requirement, and Note, forward in the section and placed them under 8.3 (d):

> *~~In addition to any contract or regulatory authority reporting requirements, t~~The organization's ~~system~~ product control process shall provide for timely reporting of delivered nonconforming product ~~that may affect reliability or safety~~.*
>
> *NOTE: Parties requiring notification of nonconforming product ~~may~~ can include suppliers, internal organizations, customers, distributors, and regulatory authorities.*

AS9100C added a new subclause entry (e) to the list of ways to deal with a nonconforming product:

> *e) by taking actions necessary to contain the effect of the nonconformity on other processes or products.*

AS9100C clarified the dispositions of "use-as-is" and "repair" and added a new Note to describe the authorized representative:

> *Dispositions of use-as-is or repair shall only be used after approval by an authorized representative of the organization responsible for the design.*
>
> *NOTE: Authorized representative includes personnel having delegated authority from the design organization.*
>
> ~~*Unless otherwise restricted in the contract, organization-designed product which is controlled via a customer specification may be dispositioned by the organization as use-as-is or repair, provided the nonconformity does not result in a departure from customer-specified requirements.*~~

AS9100C modified one requirement and deleted another one:

> *The organization shall not use dispositions of use-as-is or repair, unless specifically authorized by the customer, if ~~the product is produced to customer design, or~~ the nonconformity results in a departure from the contract requirements.*
>
> ~~*Notification shall include a clear description of the nonconformity, which includes as necessary parts affected, customer and/or organization part numbers, quantity, and date(s) delivered.*~~

The requirement below is from an earlier paragraph. No changes were made. Since it includes "subsequent actions," for example, reverification, it is appropriate for record keeping to be last in the section.

> *Records of the nature of nonconformities and any subsequent actions taken, including concessions obtained, shall be maintained (see 4.2.4)*

The deleted text below was moved to entry (d) in the list of ways to deal with nonconforming product.

> ~~*When nonconforming product is detected after delivery or use has started, the organization shall take action appropriate to the effects, or potential effects, of the nonconformity.*~~

8.4 Analysis of Data

The analysis of data provides information on customer satisfaction, conformity to product requirements, characteristics and trends of processes and products, and suppliers. The changes below were to revise a reference (from 7.2.1 to 8.2.4) and to add new references (8.2.3, 8.2.4, and 7.4).

> *The analysis of data shall provide information relating to*
>
> *a) customer satisfaction (see 8.2.1)*
>
> *b) conformity to product requirements ~~(see 7.2.1)~~ (see 8.2.4),*
>
> *c) characteristics and trends of processes and products, including opportunities for preventive action (see 8.2.3 and 8.2.4), and*
>
> *d) suppliers (see 7.4).*

8.5 Improvement

8.5.1 Continual Improvement

AS9100C added this requirement, and new Note:

> The organization shall monitor the implementation of improvement activities and evaluate the effectiveness of the results.
>
> NOTE: Continual improvement opportunities can result from lessons learned, problem resolutions, and the benchmarking of best practices.

8.5.2 Corrective Action

The requirement below switched from "cause" to "causes" to match with "nonconformities" and to be consistent with a similar sentence in 8.5.3, Preventive Action.

> *The organization shall take action to eliminate the ~~cause~~ causes of nonconformities in order to prevent recurrence.*

ISO 9000:2005 defines "review" as an activity to determine the *effectiveness* of a subject to achieve established objectives. However, the "reviewing" in the requirement below was often interpreted as checking to see if an action was taken, instead of determining its effectiveness. It has been clarified.

> *f) reviewing the effectiveness of the corrective action taken.*

AS9100C

> *g) flowing down ~~of the~~ corrective action requirements to a supplier when it is determined that the supplier is responsible for the ~~root cause~~ nonconformity , ~~and~~*
>
> *h) specific actions where timely and/or effective corrective actions are not achieved, and*

AS9100C added a new requirement to go beyond the detected problem to see if more nonconforming product might exist, and then to do something about it:

> *i) determining if additional nonconforming product exists based on the causes of the nonconformities and taking further action when required.*

8.5.3 Preventive Action

As explained under 8.5.2, Corrective Action, the "reviewing" of the action has been clarified to include determining the effectiveness of the action.

> *e) reviewing the effectiveness of the preventive action taken.*

AS9100C added a new Note to clause 8.5.3 to give examples of preventive action opportunities:

> *NOTE: Examples of preventive action opportunities include risk management, error proofing, failure mode and effect analysis (FMEA), and information on product problems reported by external sources.*

CONCLUSION

We hope that this book has provided many perspectives to guide you through planning, implementing, or improving your quality management system. Through decades of cumulative aerospace, aeronautics, and space quality management systems and continuous improvement experience, we have learned what works and had a sincere desire to communicate that to you, the reader. Successful companies in these industries today and in the future need to compete globally—only those with efficient systems will prevail.

Continual reference to this guide will enable you (or your team) to capture new ideas as you become more and more familiar with the AS9100 standard and what is contained in these pages. Plan well, train your core team, and implement for the right reasons—to be the best—not just because it is a customer requirement. Lack of ownership and understanding by management of the benefits of following AS9100 to the level suggested in this book is relevant within the industry. You and your organization have a grand opportunity to stand above the rest by truly working to the intent of AS9100.

Best wishes in your continual improvement endeavors!

AS9100 DOCUMENTED REQUIREMENTS BY SECTION

Sec	Title	Description	Type
4.2.2	Quality manual	Quality manual	M
4.2.3	Control of documents	Documented procedure	P
4.2.4	Control of records	Documented procedure	P
8.2.2	Internal audit	Documented procedure	P
8.3	Control of nonconforming product	Documented procedure	P
8.5.2	Corrective action	Documented procedure	P
8.5.3	Preventive action	Documented procedure	P
5.6.1	General	Management reviews	R
6.2.2(e)	Competence, awareness, and training	Employee skills	R
7.1(d)	Planning of product realization	Product fulfillment	R
7.2.2	Review of requirements related to the product	Requirements review	R
7.3.2	Design and development inputs	Design inputs	R
7.3.4	Design and development review	Design reviews	R
7.3.5	Design and development verification	Design verification	R
7.3.6	Design and development validation	Design validation	R
7.3.7	Control of design and development changes	Design changes	R
7.4.1	Purchasing process	Supplier evaluation	R
7.5.2(d)	Validation of processes for production and service provision	Process validation	R
7.5.3	Identification and traceability	Product identification	R
7.5.4	Customer property	Customer product review	R
7.6 (a)	Control of monitoring and measuring devices	Calibration standards	R
7.6	Control of monitoring and measuring devices	Previous results	R
7.6	Control of monitoring and measuring devices	Results of calibration	R
8.2.2	Internal audit	Audit results	R
8.2.4	Monitoring and measurement of product	Product conformance	R
8.3	Control of nonconforming product	Nonconforming nature	R
8.5.2(e)	Corrective action	Corrective action results	R
8.5.3(d)	Preventive action	Preventive action taken	R

Legend: M = manual / P = procedure / R = record

GLOSSARY

C_p, C_{pk} (capability index)—Statistical measure of a process being in control or equipment being precise and accurate, calculated using the standard deviation of the sample. C_p is calculated using the standard deviation of the sample and measures the overall capability. C_{pk} is the measure of a process being centered against the mean of the sample.

CTC (critical to cost)—A measure or characteristic deemed important in controlling the cost of manufacture.

CTQ (critical to quality)—A measure or characteristic deemed of importance by the customer or manufacturer.

DFSS (design for Six Sigma)—A methodology for creating robust designs utilizing Six Sigma tools.

DMAIC (define–measure–analyze–innovate/improve/implement–control)— The DMAIC project methodology has five phases: Define the problem, the voice of the customer, and the project goals specifically. Measure key aspects of the current process and collect relevant data. Analyze the data to investigate and verify cause-and-effect relationships. Determine what the relationships are, and attempt to ensure that all factors have been considered. Seek out root cause of the defect under investigation. Improve or optimize the current process based on data analysis using techniques such as design of experiments, poka-yoke, or mistake-proofing, and standard work to create a new, future-state process. Set up pilot runs to establish process capability. Control the future state process to ensure that any deviations from target are corrected before they result in defects. Control systems are implemented such as statistical process control, production boards, and visual workplaces, and the process is continuously monitored.

FAA (Federal Aviation Administration)—Governing body that relays requirements for design, manufacture, and repair and overhaul of aerospace and avionic equipment.

5S (sort, set in order, shine, standardize, sustain)—A lean foundational principle that stipulates keeping only what is needed for the job in the work area and that those items are easily available to the person performing the task.

FOD—1) foreign object detection—Method to ensure that no foreign material has inadvertently entered the part. 2) foreign object damage—Damage that occurs when foreign material affecting the safe function of the product has not been prevented or detected.

hoshin (Japanese—roughly translates to "policy deployment")—A strategic planning method of cascading the major business objectives and strategies throughout the organization, helping everyone to focus tactical work on common goals and be accountable to such.

layout—A lean principle that encourages the simplest flow of product or code. The goal is to reduce or eliminate time taken to transport product or for information to travel from one step to the next.

lean—A methodology initially developed from Deming principles by Toyota (Toyota Production System) that simplifies and mistake-proofs process and procedure, thereby creating a physical process control in administrative and manufacturing environments.

QFD (quality function deployment)—A method within DFSS and Six Sigma to translate the voice of the customer (VOC) into functional requirements. Can be used in software and product design.

QMS (quality management system)—Prior to ISO 9001:2008 called a quality system.

RTA (requirement traceability to test acceptance)—A matrix that documents business requirements to functional specifications, then technical specifications, then acceptance criteria to ensure that final software or hardware meets customer specifications.

Six Sigma—Follows the DMAIC methodology for process improvement and control utilizing Six Sigma tools.

SPC (statistical process control)—Utilization of statistical tools, such as control charts and C_p studies, to measure the stability of a process.

visual—A lean principle that promotes making the workplace, documentation, or code easily discernible, thereby mistake-proofing the environment, use of correct documents, or code structures.

REFERENCES

AS9100 Quality Management Systems—Aerospace Requirements

AS9102: First Article Inspection—Aerospace Requirements

AS9103: Variation Management of Key Characteristics

AS9101: Quality Management Systems Assessment

AS9003: Inspection and Test Quality System

Bibliography

The Bibliography for AS9100C has been updated to reflect new standards, new editions of standards, and withdrawn standards since the publication of AS9100B.

~~*AS/EN/SJAC 9102 Aerospace First Article Inspection Requirement*~~

AS/EN 9110 Quality Management Systems—Requirements for Aviation Maintenance Organizations.

AS/EN 9120 Quality Management Systems—Requirements for Aviation, Space and Defense Distributors.

ISO 9000~~:2000~~ Quality management systems—Fundamentals and vocabulary.

ISO 9001~~:2000~~ Quality management systems—Requirements.

ISO 9004:~~2000 Quality management systems - Guidelines for performance improvements~~ Managing for the sustained success of an organization—A quality management approach.

ISO 10007~~:1995~~ Quality management systems—Guidelines for configuration management

ISO 19011~~:2002~~ Guidelines for quality and/or environmental management systems auditing

Informational Web sites:

- http://www.iso.ch
- http://www.asq.org
- http://www.qualitydigest.com
- http://www.informintl.com/home/
- http://www.moorhill.com

Also by Erik Myhrberg:

1. *A Practical Field Guide for ISO 9001:2008* (Milwaukee: ASQ Quality Press, 2009).
2. *The ISO 9001:2008 QMS "Bluesheet"* (Phoenix: Moorhill Publishing, 2008).
3. *The ISO 14001:2004 EMS "Greensheet"* (Phoenix: Moorhill Publishing, 2005).
4. *The 18001:2007 HSMS "Redsheet"* (Phoenix: Moorhill Publishing, 2007).
5. *The ISO Core (C^4) Four Pocket Guide* (Phoenix: Moorhill Publishing, 2002).
6. *The ISO 27001:2005 ISMS "Goldsheet"* (Phoenix: Moorhill Publishing, 2006).

ASQ

ASQ Membership

Research shows that people who join associations experience increased job satisfaction, earn more, and are generally happier*. ASQ membership can help you achieve this while providing the tools you need to be successful in your industry and to distinguish yourself from your competition. So why wouldn't you want to be a part of ASQ?

Networking

Have the opportunity to meet, communicate, and collaborate with your peers within the quality community through conferences and local ASQ section meetings, ASQ forums or divisions, ASQ Communities of Quality discussion boards, and more.

Professional Development

Access a wide variety of professional development tools such as books, training, and certifications at a discounted price. Also, ASQ certifications and the ASQ Career Center help enhance your quality knowledge and take your career to the next level.

Solutions

Find answers to all your quality problems, big and small, with ASQ's Knowledge Center, mentoring program, various e-newsletters, *Quality Progress* magazine, and industry-specific products.

Access to Information

Learn classic and current quality principles and theories in ASQ's Quality Information Center (QIC), *ASQ Weekly* e-newsletter, and product offerings.

Advocacy Programs

ASQ helps create a better community, government, and world through initiatives that include social responsibility, Washington advocacy, and Community Good Works.

Visit www.asq.org/membership for more information on ASQ membership.

*2008, The William E. Smith Institute for Association Research

ASQ Certification

ASQ certification is formal recognition by ASQ that an individual has demonstrated a proficiency within, and comprehension of, a specified body of knowledge at a point in time. Nearly 150,000 certifications have been issued. ASQ has members in more than 100 countries, in all industries, and in all cultures. ASQ certification is internationally accepted and recognized.

Benefits to the Individual

- New skills gained and proficiency upgraded
- Investment in your career
- Mark of technical excellence
- Assurance that you are current with emerging technologies
- Discriminator in the marketplace
- Certified professionals earn more than their uncertified counterparts
- Certification is endorsed by more than 125 companies

Benefits to the Organization

- Investment in the company's future
- Certified individuals can perfect and share new techniques in the workplace
- Certified staff are knowledgeable and able to assure product and service quality

Quality is a global concept. It spans borders, cultures, and languages. No matter what country your customers live in or what language they speak, they demand quality products and services. You and your organization also benefit from quality tools and practices. Acquire the knowledge to position yourself and your organization ahead of your competition.

Certifications Include

- Biomedical Auditor – CBA
- Calibration Technician – CCT
- HACCP Auditor – CHA
- Pharmaceutical GMP Professional – CPGP
- Quality Inspector – CQI
- Quality Auditor – CQA
- Quality Engineer – CQE
- Quality Improvement Associate – CQIA
- Quality Technician – CQT
- Quality Process Analyst – CQPA
- Reliability Engineer – CRE
- Six Sigma Black Belt – CSSBB
- Six Sigma Green Belt – CSSGB
- Software Quality Engineer – CSQE
- Manager of Quality/Organizational Excellence – CMQ/OE

Visit www.asq.org/certification to apply today!

ASQ